African Mind, Culture, and Technology

Yamikani Ndasauka

African Mind, Culture, and Technology

Philosophical Perspectives

Yamikani Ndasauka ⓘ
Philosophy Department
University of Malawi
Zomba, Malawi

ISBN 978-3-031-62978-5   ISBN 978-3-031-62979-2   (eBook)
https://doi.org/10.1007/978-3-031-62979-2

This Palgrave Macmillan imprint is published by the registered company Springer Nature Switzerland AG.
The registered company address is: Gewerbestrasse 11, 6330 Cham, Switzerland

Paper in this product is recyclable.

PREFACE

As a scholar and philosopher from Africa, I have long been fascinated by the relationship among African knowledge systems, cultural traditions, and the rapid pace of technological change sweeping across our continent. This book, *African Mind, Culture, and Technology: Philosophical Perspectives*, represents the culmination of my intellectual journey to interrogate these issues and to articulate a distinctly African vision for shaping more inclusive, culturally resonant, and ethically grounded technological futures.

My motivation for writing this book comes from a sense of responsibility to contribute to the growing discourse on the role of African philosophy in guiding innovation and development on the continent. Throughout my academic career, I have been struck by the relative paucity of African voices and perspectives in global debates around science, technology, and society. I have often encountered narratives portraying Africa as a passive recipient of imported technological solutions rather than a vibrant source of indigenous creativity and knowledge production. This book is my attempt to challenge these stereotypes and to foreground the rich tapestry of African thought that has long grappled with questions of epistemic justice, cultural identity, and ecological sustainability in the face of modernity's transformative forces.

I am particularly inspired by the humanistic values and communitarian ethics that underpin many African worldviews. The concept of Ubuntu, which emphasises the intrinsic dignity and interconnectedness of all beings, has been a guiding light for my intellectual journey. In a world increasingly dominated by individualistic and instrumentalist logic, I believe that Ubuntu offers a powerful framework for re-imagining the relationship between technology and society in more holistic, compassionate, and life-affirming ways. Throughout this book, I draw on Ubuntu and other African philosophical traditions to propose alternative paradigms of innovation that prioritise collective well-being, social solidarity, and environmental stewardship over mere economic growth or efficiency as the essence of technology.

My inspiration for this project also stems from my own lived experiences as an African researcher navigating the terrain of knowledge production in a globalised academy. As a scholar of philosophy of technology, I have had the opportunity to immerse myself in debates around the social and ethical implications of emerging technologies. However, I often found myself grappling with the limitations of Western philosophical frameworks in capturing the nuances and complexities of African realities. This book is my attempt to bridge this gap by bringing African epistemologies, ontologies, and axiologies into dialogue with global discourses on technology and society.

Throughout my research career, I have been fortunate to collaborate with a diverse network of scholars and practitioners across Africa and beyond who share my commitment to decolonising knowledge production and amplifying marginalised voices in the academy. I have also been inspired by African communities' resilience, creativity, and wisdom in navigating the challenges and opportunities of technological change. This book is a testament to these communities' collective insights and aspirations and the need for more inclusive and participatory approaches to innovation that centre their agency.

I hope this book will catalyse further research, dialogue, and action on the intersections of African philosophy, culture, and technology. I believe that by taking the insights and provocations of African thinkers seriously, we can chart alternative pathways of innovation that are more attuned to African communities' needs, values, and aspirations. This is not a call for isolation or exceptionalism but rather an invitation to engage in more authentic and reciprocal forms of knowledge exchange across cultural and disciplinary boundaries. As we grapple with the existential challenges of our time, from climate change to rising inequality to the unchecked power of technology, I believe that African wisdom offers vital resources for re-imagining the future in more holistic, relational, and regenerative ways.

Zomba, Malawi   Yamikani Ndasauka

CONTENTS

LIST OF FIGURES

CHAPTER 1

Why African Philosophy of Technology?

1.1 INTRODUCTION

Technology is advancing at lightning speed across our beloved continent of Africa, bringing immense possibility and risk. From artificial intelligence to biotechnology to blockchain, we are seeing all kinds of gadgets and systems taking hold in our economies and societies, influencing all aspects of life. This accelerating wave of techno-social change requires us to think carefully about how these technologies align with time-honoured African values and ways of thinking. But so much of the conversation about technology these days reflects assumptions that come out of the Western worldview—assumptions about progress, efficiency, and individualism that often leave marginalised communities out in the cold. And the business models driving innovation seem laser-focused on profits and shareholders rather than addressing real African problems. Even technologies designed for Western consumer markets sometimes fail to meet local African communities' real needs and priorities. These impacts reflect definite biases that empower connected urban elites while excluding rural villages and informal settlements. We consistently see women, youth, elders, and people with disabilities facing barriers to accessing and helping to design technologies. At the same time, automation threatens the livelihoods of the masses without any social safety net to fall back on. As such, Africa stands at a crossroads with all this tech change happening; we have huge opportunities but also huge risks to navigate.

Y. Ndasauka, *African Mind, Culture, and Technology*, https://doi.org/10.1007/978-3-031-62979-2_1

Despite all these technological changes, Africa has yet to truly tap into the holistic wisdom and resources within our rich cultural traditions—from the Ubuntu philosophy of our interconnectedness to long-standing practices of spirited innovation grounded in community. Our African value systems that emphasise social ethics, communal ties, and living in balance with nature contain vital guidance that could steer technology towards advancing society as a whole rather than feeding individualistic greed. Centuries before contact with Europe and the tragedy of colonisation, great African civilisations had already innovated sophisticated tools ranging from architecture to medicine to metallurgy (Hasenöhrl, 2021). Tools that demonstrated advanced technical capabilities were later systematically erased from mainstream histories. Truly realising the emancipatory potential of technology requires us to invest deeply in African-centred scholarship that illustrates our own perspectives on technology and society. As I will show, tapping into the holistic wisdom of our ancestors and traditions remains vital if we want technology to uplift our communities rather than divide them.

Sadly, the African philosophy of technology currently constitutes a significant knowledge-production gap. The field remains minimally developed, with few scholars explicitly engaging technology through African epistemic frameworks. Reasons include the dominance of Eurocentric paradigms that long dismissed African philosophy outright. When African philosophy finally gained recognition, the focus emphasised recapturing past belief systems rather than tackling contemporary problems. With philosophy marginalised in many African universities, specialised subdisciplines like the philosophy of technology have had little chance to take root. Scholars have just begun examining fields like indigenous innovation and applying Ubuntu to emerging technologies. However, such contributions remain highly fragmented and limited in scope.

As I will show in this chapter, this book will act as an inspiration for growing the African philosophy of technology into a thriving field. Advancing Africa's technical heritage with the hope of overturning dominant assumptions of pre-colonial stagnation is vital and arguably the way to push the African agenda on technology forward. Further, social constructivism demonstrates technology's embeddedness in cultural values and social power dynamics rather than following universal technical dictates. This critiques diffusion patterns that pronounce biases of concentrated expertise and benefits among elites. In contrast, principles of Ubuntu ethics are applied to guide technology in humanistic directions to benefit communities. This book contributes to the conceptual foundations and pragmatic pathways to grow the African philosophy of technology into a vibrant field capable of steering science, engineering, and innovation towards just, empowering, and sustainable futures across the world's most demographically dynamic continent.

The chapter is structured as follows. The first section reviews factors explaining the field's minimal development, including the marginalisation of African philosophy, holistic traditional worldviews, lack of focus on contemporary issues, and deficient institutions and networks. Then, I explore how Ubuntu and communal values offer guidance, though contributions are currently limited in scale and fragmented. The following section argues for countering mainstream discourse biases, foregrounding marginalised voices, resisting universalisms, and grounding technology in African communalism. Then, I provide vital new insights, further guide research developing African philosophy of technology, warn against homogenising diverse African cultures under singular notions of 'African philosophy', and instead advocate contextual plural philosophising. Finally, I summarise the book's expansive examination of technology through historical, ethical, cultural, policy and innovation lenses to articulate humanistic visions guided by African thought systems.

1.2 Lack of Scholarship in African Philosophy of Technology

The scholarship on African philosophy of technology is still minimal, representing a significant knowledge gap in this field. Despite some valuable contributions, the literature remains small and fragmented. Several factors help explain this lack of development. First, the field of African philosophy itself struggled for recognition until recently. Western philosophers long discounted the idea that African philosophy existed, claiming that rationality and abstraction were foreign to African cultures (Hallen, 2002). It was not until the mid-twentieth century that philosophers like Placide Tempels, Kwame Gyekye, and Kwame Nkrumah rigorously articulated the metaphysical frameworks, epistemologies, and value systems within African worldviews. They demonstrated that rigorous philosophical discourse was present in Africa. With African philosophy only now growing into a mainstream discipline, subfields like the philosophy of technology have had little chance to take root.

The lack of engagement also stems from the nature of traditional African knowledge systems. African worldviews tended to integrate spirituality, ethics, and an emphasis on communal relationships rather than divide domains into specialised subdisciplines. With knowledge more incorporated into everyday life, there was less impetus to develop systematic philosophical study of narrow subject areas. The Western analytic tradition, in contrast, prized precise, logical analysis of narrowly defined problems (Hallen, 2002). So, the specialised abstract style of philosophy of technology did not organically emerge in Africa before extensive Western contact.

Further, scholarship on African philosophy has primarily focused on recapturing traditional thought, particularly pre-colonial traditions that were marginalised by colonialism (Serequeberhan, 1994). While this is an important project, the emphasis on past perspectives has overshadowed the examination of how African philosophy could critically engage with contemporary techno-social issues. This focus on reclaiming traditional belief systems has come at the expense of applying African philosophical frameworks to pressing modern challenges, such as the rapid proliferation of technology across the continent. By concentrating efforts on recuperating ancient wisdom, African philosophers have had limited engagement with the ethical, social, and existential implications of emerging technologies like artificial intelligence, biotechnology, and digital platforms.

The lack of institutional support further compounds the gaps in African philosophical engagement with contemporary techno-social issues. The text highlights that the humanities, in general, and philosophy specifically, remain under-prioritised in many African universities. This marginalisation of philosophical inquiry within academic institutions has limited the development of specialised subfields, such as the philosophy of technology. The absence of robust institutional structures and resources dedicated to nurturing philosophical research has inhibited the advancement of scholarship in these areas. Without strong departmental support, funding, and intellectual communities focused on exploring the intersections of technology, society, and African thought, philosophers have found it challenging to pursue sustained and impactful work in this domain.

That said, some valuable contributions to the African philosophy of technology have emerged. Scholars like Lauer (2017) have suggested that African knowledge systems contain rich resources for ontological and epistemological insights about the nature of technology and its relationship to society. For example, Afrocentric understandings of Ubuntu could inform philosophical debates about the ethical implications of advanced AI and robotics (Gwagwa et al., 2022; Jecker et al., 2022). Additionally, communal values in African cultures could ground philosophical perspectives on directing technology towards collective needs rather than individualistic greed (Verharen et al., 2014). Kaschula and Mostert (2011) have also discussed the impacts of communication technologies on African oral traditions and knowledge-sharing practices. They have explored issues like digitisation, changing media ecosystems, and the movement of folklore into digital spaces and their cultural implications. Others have examined overlaps between artistic, spiritual, and technical dimensions in African material cultures for insights into the philosophy of techno-social systems (Oladumiye, 2014).

While these contributions provide initial foundations, this body of literature remains highly limited in scope, scale, and coherence. This entails a lack of systematic philosophical analysis of technology's core epistemic, metaphysical, ethical, and socio-political facets within African knowledge frameworks. Very few thinkers have taken up that challenge. In a continent with over a billion people, the small scholarly community explicitly engaged with the African philosophy of technology severely limits perspective diversity, critical discourse, and knowledge accumulation. In addition, the literature lacks robust region-wide exchanges, with most work emerging within narrow academic circles in South Africa, Nigeria, and a few other centres. Yet, technology offers an urgent concern for the African mind and culture, requiring pan-African philosophical discussions.

The emergent field of African philosophy of technology currently constitutes a substantial knowledge gap in need of scholarly engagement. Given the growing role of technology across Africa, developing rigorous philosophical perspectives rooted in African worldviews, values, and epistemologies can provide essential local and global insights. However, realising this potential will require investing more resources and expanding capacity to nurture a broader community of thinkers and exchanges across the continent. The current marginal state of scholarship represents a missed opportunity in global philosophical discourse. Filling this gap will be essential for bringing African perspectives to the fore in ethics, policy, and practice regarding technological change across Africa in an era of digitisation.

1.3   NEED FOR AN AFRICAN PHILOSOPHY
OF TECHNOLOGY

The exponential growth of technology across Africa demands philosophical engagement grounded in African knowledge frameworks. However, most technology designs and policies rest on individualistic assumptions that conflict with traditional African values centred on communal interdependence. Mainstream Western philosophy of technology often promotes values of efficiency, productivity, hyper-individualism, and technological solutionism in ways misaligned with African cultures. For example, technology designed within a capitalist framework can undermine social solidarity and reciprocity. A philosophy of technology grounded in African communal thinking provides an alternative foundation to guide technology's trajectory in more humane directions.

African communalism emphasises that individual identity and flourishing depend on harmonious social relationships (Mbiti, 1970; Menkiti, 1984). This differs from the individualism prevalent in Western liberal thought that often informs technological design and economics. Philosophy centred on African communalism would critique technical systems that disrupt social cohesion or undermine people's duties to their communities. An African philosophical perspective grounded in communal values would highlight the need to structure such technologies to foster unity and conciliation instead. Likewise, automation that destroys livelihoods and communities in the name of efficiency would face moral condemnation. The goal for Africa and Africa's philosophy of technology would be to apply technology to nourish social solidarity. This communal orientation aligns with the African ethical concept of Ubuntu, which recognises that our humanity is bound up with the humanity of others (Tutu, 1999). An African philosophy of technology informed by Ubuntu would ask: Does this technology affirm our shared dignity and interdependence? Or does it isolate people and erode communal ties? Tools that undermine Ubuntu would face deep scepticism. An African communal philosophy of technology would also emphasise collective benefit over individual profit. Technological design often caters to those with purchasing power rather than considering marginalised populations. In contrast, a philosophy prioritising community would develop technologies tailored to affordability, accessibility, and social inclusion. It would value traditional knowledge and indigenous innovation as much as Western tech.

For us to realise such a communally grounded African philosophy of technology, we would require expanding scholarship and engaging African knowledge systems to address contemporary techno-social dilemmas. It would mean building new Afrocentric frameworks and analysing how tools interface with culture and society. It necessitates technology research, development, and policymaking that cuts against profit-driven individualism with people-centred communal thinking. However, articulating a holistic African communal philosophy of technology requires far more scholars engaging in this discourse across disciplines and institutions continent-wide. It means drawing on epistemic resources from traditional knowledge systems and modern African philosophy. Moreover, it requires technology practitioners to be open to alternative visions that break with Western individualistic assumptions.

1.4 UTILISING AFROCENTRIC, POSTCOLONIAL, AND AFRICAN PHILOSOPHICAL FRAMEWORKS

Mainstream discourses around technology and society overwhelmingly stem from Western epistemological foundations. These paradigms have been accused of being rooted in coloniality and capitalism. Technology designed within this paradigm tends to reflect and exacerbate inequities already ingrained in social structures and institutions. Mainstream technology discourses have often ignored indigenous perspectives and non-Western systems of thought that could provide vital ethical insights and alternative visions. Centring African and decolonised philosophical frameworks helps counterbalance deficient mainstream views in several essential ways. First, it foregrounds marginalised voices and knowledge excluded from technology debates. Philosophy grounded in the lived experiences of African and postcolonial peoples surfaces urgent critical concerns otherwise dismissed by privileged Western thinkers. It demands reckoning with inequality and justice issues elided in conventional tech discourse. Second, African and postcolonial thought provides metaphysical, ethical, and epistemological insights with immense significance for technology not found in Western philosophy. These concepts offer visions of collective dignity, while Afrocentric ideas of harmony with nature contrast technology's tendencies towards exploitation and domination (Verharen et al., 2014).

Additionally, African values counter hyper-individualism prevalent in Western thought and technology design. African humanism helps re-centre technology debates on humanistic goals rather than profits. Engaging African philosophy can help shift technology's trajectory towards more collective, responsible, sustainable futures. Furthermore, African and Afrocentric thought resist the imposed universalisms that erase pluralism and ignore local contexts. We have seen technologies that are designed as a one-size-fits-all solution within a monolithic paradigm generally fail communities in the Global South. African philosophy insists on foregrounding plural perspectives and situated nuanced technology applications. Likewise, traditions of African critical theory illustrate issues of power, inequality, and ideological hegemony intertwined with technology in ways Western philosophy frequently ignores. Moreover, Afrocentric frameworks emphasise centring marginalised peoples as subjects, not just objects, in technology discourses. This book will show that this orientation is vital for ethical, inclusive tech debates and policies.

Examining technology via African cosmologies highlights metaphysical dimensions severed in Western technoscientific worldviews. Reconnecting technology and spirituality counters unreflective rationalism and instrumentalism. African insights also question notions of technology as neutral or deterministic, showing how tools interact dynamically with social forces and human values. Some existing works demonstrate how African and Afrocentric philosophical frameworks can productively analyse technology. For example, accounts of African indigenous innovation systems highlight how tools were historically developed to promote communal welfare, in contrast to capitalistic systems (Dei, 2020).

With advanced technologies increasingly mediating broad aspects of African economic, political, social, and cultural life, developing philosophical frameworks grounded in African epistemic traditions is needed. Only philosophical discourse rooted in Africa's knowledge heritage can help guide technology's further diffusion throughout the region in ethically and socially just directions. We should offer an African technology philosophy grounded in communal values. Decolonised thought is essential for humanising technological change and steering practitioners and policymakers away from exploitative paradigms. Realising such a field requires platforms and programmes dedicated to enhancing Afrocentric, postcolonial, and African philosophical engagements with technology throughout educational institutions and public spheres across the continent.

1.5 THE PROBLEM OF GENERALISATION IN AFRICAN PHILOSOPHY

Now, let us turn our discussion to African philosophy. I will systematically discuss its tenets in Chap. 5. For now, I want us to consider the debate on the breadth of African philosophy vis-à-vis the African people and continent. The field of African philosophy has often struggled with homogenising diverse African cultures and knowledge systems under a singular umbrella. Any attempts to articulate an essentialist 'African philosophy' risk erasing differences in values, beliefs, and worldviews across the vast continent. The impulse to homogenise stems from the marginalisation of African philosophy within Eurocentric paradigms that dismissed the possibility of advanced abstract thought outside the West. Early African philosophers thus sought to prove the existence of coherent philosophical systems codified within African cultures to counter Western biases. This produced overly broad generalisations about 'African thought'. However, lived philosophies in Africa varied widely between specific ethnic groups, regions, and historical eras. No universal archetypal African worldview exists. For example, the humanist values underpinning Ubuntu among Nguni Bantu peoples of Southern Africa differ from the vitalism and divine kingship beliefs that shaped the Akan of West Africa. Diversity ranges from Islamic philosophies in North Africa to the spiritualities of the San hunter-gatherers in Southern Africa.

Essentialising these multiplicities into a singular 'African philosophy' risks projecting particular localised customs or decontextualised principles as universally representative of all Africans, universalism which this book seeks to dispel. This elides differences in how diverse African peoples produced meaning, truth, and value from their distinct historical and material contexts. This tendency towards homogenisation also commonly privileged educated male perspectives, silencing women's voices and non-elite modes of knowledge production. Additionally, essentialist discourse about 'African philosophy' has often focused excessively on recapturing pre-colonial belief systems in the past rather than engaging contemporary problems. These issues have led many thinkers to critique appeals to a monolithic African philosophy as empirically dubious and politically problematic. Joseph Omoregbe (1998) argued that philosophical systems within Africa are plural, not uniform. Tsenay Serequeberhan (1994) similarly concluded that seeking an 'essential African mind' is untenable.

The solution, they suggest, lies in phasing out homogenising talk of 'African philosophy' in favour of context-conscious analysis grounded in particular African realities and exchanges between them. This means philosophers in lived African spaces engage philosophically with those situated contexts and cosmologies. This approach accommodates diversity and dynamism rather than assuming universality or static traditions. It foregrounds plural perspectives in dialogue, not homogeneous systems lineages. It stresses emergent, unfinished philosophising embedded in everyday African lives, not just colonial anthropological artefacts or ancient beliefs fixed in a mythic past. Developing an inclusive African philosophy demands dropping imposed abstractions and listening to marginalised philosophical voices speaking from and into their authentic situations. This grounds ethical, metaphysical, and social analysis in real material contexts to avoid decontextualised, inaccurate generalisations. Academics alone cannot define such an African philosophy; it requires weaving together perspectives from diverse publics and views across the continent.

This particularising contextual approach is evident in various contemporary schools like ethnophilosophy, sage philosophy, and political/ideological philosophy that avoid homogenising assumptions. It is also central to networks like the Council for the Development of Social Science Research in Africa (CODESRIA) that facilitate philosophical dialogues between local African contexts. Some also see promise in analytical perspectives that rigorously probe concepts within African languages and meaning systems. From this view, 'genuine' African philosophy will only gradually emerge through such ground-up, grassroots philosophising by and between diverse African peoples speaking from their unique standpoints. Attempts to construct unified doctrines a priori using decontextualised abstractions will collapse. The terrain of African philosophy must be mapped philosophically from the inside to render an authentic portrait reflecting varied landscapes. Of course, the risk of extreme relativism is real here. However, ethical humanism emphasised by many African philosophers may provide universals to orient particularised discourse. Respectful integrative work identifying specific shared concerns across situations remains possible, though complex, given vast diversities. The key is avoiding imposed generalisations detached from lived realities.

While it is essential to acknowledge that Africa is not homogenous, as presented above and that the continent encompasses diverse perspectives and value systems, in discussing the ideas in this book, I intend not to imply that these singular viewpoints encapsulate the full range of African thought. The philosophies explored here represent influential strands of discourse within Africa but do not constitute the entirety of 'African philosophy'. As critics have highlighted, given the continent's vastness, it would be impossible for any text to capture African thought's totality. However, by delving deeply into certain prominent schools of thought, we can gain insight into influential ways of thinking that have shaped discourse and culture across parts of Africa. Though not universally representative, these ideas provide windows into major philosophical trends and reveal the complexity of African traditions. There are always multiple lenses to view a subject as vast as an entire continent. In presenting this snapshot, the book's aim is not generalisation but rather an illustration of specific critical threads of African philosophy that may significantly impact our understanding and conception of technology. Moreover, we are African Philosophers, and though debating through different frameworks, contexts and epistemologies, we are generally perceived as speaking on behalf of Africa.

1.6   Shaping Just Technology for Africa

Africa finds itself amid a digital revolution, with technologies proliferating across economies and societies at unprecedented rates. However, this rapid technologisation risks exacerbating inequities and harm if not approached thoughtfully. Fostering widespread philosophical discussion on shaping technology according to principles of justice is critical for realising technology's emancipatory potential in Africa. Mainstream technology discourse ignores ethics and justice, focusing narrowly on efficiency and productivity under capitalist paradigms. As we have noted, technology designed by and for wealthy corporations and consumers often impoverishes and disadvantages marginalised groups. The values and assumptions driving technological innovation are unjust, let alone the impacts. Now more than ever, it is imperative to generate broad public dialogue to articulate alternative visions for just technology. This entails critically examining the politics and privilege inherent in mainstream technical systems and re-imagining technology centred on the needs of ordinary Africans, especially the vulnerable and oppressed.

Critical philosophical questions to spur these discussions include but are not limited to the following:

1. How can we ensure technologies affirm the dignity and humanity of all people?
2. How can innovation be directed to enhancing equity and social welfare rather than exacerbating inequality?
3. How can we shape tools to empower marginalised groups rather than further disadvantage them?
4. How can we build participatory technology systems that ordinary citizens can meaningfully influence?
5. How can we develop technical systems reflecting plural values and local contexts across Africa's diversity rather than imposed universalist assumptions?

This book attempts to address some of these questions. However, fully addressing them requires dialogues among the African public, developers, policymakers, and civil society. It means making technology discourse participatory by teaching philosophical perspectives on technology and justice in schools and universities. It necessitates scholars applying African philosophical concepts like Ubuntu to articulate alternative paradigms for socially just technology. Indeed, just technology requires radically rethinking the egoism and shareholder primacy driving most current systems. African communalism and mutual care traditions could ground more ethical technical systems prioritising collective advancement and solidarity over individualism and profits. Further, indigenous innovation practices and emerging Afrocentric design approaches also offer foundations to build.

1.7   Scope of the Book

This book provides a comprehensive, expansive, and far-reaching philosophical examination of technology in Africa, synthesising insights from history, ethics, culture, policy, and innovation studies. The overarching goal is to articulate conceptual frameworks rooted in African thought traditions to guide empowering, humanistic technology innovation across the vast, diverse continent. The scope ranges from excavating Africa's rich indigenous innovation heritage to applying value systems like Ubuntu to tackling complex dilemmas wrought by emerging, rapidly proliferating technologies. A central thread demonstrates how African philosophies of communalism, social ethics, spirituality, and ecological harmony contain meaningful resources for shaping just, equitable technological futures, countering individualistic, dehumanising mainstream paradigms. The primary goal of this book is to dispel misleading assumptions that Africa lacks indigenous technical innovation or capabilities. The book highlights the creativity evident across the continent before colonisation in architecture, medicine, and metallurgy. This evidence, I hope, overturns dominant narratives that erase or downplay Africa's rich homegrown innovation heritage. Additionally, the book challenges technological determinism and instrumentalism perspectives, which view technology as an autonomous, value-neutral force. Social constructivism demonstrates technology's embeddedness within cultural contexts, power dynamics, and value systems.

This book also aims to expose deep biases within mainstream technology discourse and innovation models that systematically exclude African voices and perspectives. Much of the current technology design ignores African needs, values, and realities. By advancing African knowledge systems and ethics, the book helps amplify marginalised African voices to articulate empowering, humanistic visions for technology. Bringing African philosophies like Ubuntu into technology debates provides vital ethical guidance from the continent's intellectual traditions. The book also proposes strategies to substantially increase indigenous innovation capabilities and African participation. This expands local control over technology design and integration. The book advocates significant reforms to decision-making structures, educational systems, and funding programmes that can nurture appropriate, empowering technical solutions tailored to African contexts. Democratising innovation requires building endogenous capacity in science, engineering, and technology policy. Finally, the book supplies extensive philosophical and practical principles to guide ethical, humanistic technology design and to construct robust policy frameworks supporting just, sustainable technology innovation in Africa. The book articulates strategies for applying African wisdom traditions to empower African societies to direct technology's trajectory on their terms to promote human development.

The book is structured as follows. Chapter 2 aims to overturn dominant, misleading narratives that systematically erase Africa's technical ingenuity and establish strong conceptual foundations. Profiling fields like metallurgy, architecture, and medicine reveal advanced, sophisticated innovations that thrived before colonisation. This grounds contemporary innovation in Africa's creative, original philosophies rather than importing foreign models. Chapter 3 analyses the socio-technical factors influencing the adoption and integration of technologies in Africa. It examines how digital platforms, AI, and other innovations spread unevenly, creating digital divides and development gaps between groups based on gender, geography, age, and class. Chapter 4 examines the complex intersections of modern technologies with diverse dimensions of African culture, including oral traditions, languages, artistic expression, social institutions, and communal practices. It argues against simplistic assumptions of technology as a unidirectional force displacing traditional forms, instead revealing nuanced narratives of creative adaptation alongside problematic disruption.

Chapter 5 constructs an African philosophy of technology guided by principles of community, sacred ecology, and care rather than technical rationalism. An Afrocentric framework envisions technologies designed to harmonise with, not dominate, nature and society. Chapter 6 develops social constructivist perspectives, demonstrating technology's embeddedness in social contexts, values, and power dynamics rather than following universal technical logic. Diffusion patterns exhibit demographic biases and uneven capability constraints. Impacts are shown to be ambiguous, requiring ethical guidance. This analysis frames technology as sociocultural artefacts to be consciously directed, not as an autonomous force. Chapter 7 proposes Ubuntu's communitarian ethics as vital touchstones to assess and steer technology trajectories in collectivist, humanistic directions, resonating with African moral thought. Chapter 8 explores the interplay between African spirituality and technological innovation, examining how traditional African beliefs, practices, and values shape responses to the promises and perils of emerging technologies. Finally, Chap. 9 synthesises key insights to propose frameworks and practical strategies for shaping technology to advance human flourishing in Africa. It outlines principles grounded in African ethics to guide innovation policies and practices, emphasising the importance of decolonising technology, building endogenous capabilities, and promoting participatory design.

This book builds on emerging interdisciplinary scholarship at the intersection of African studies, philosophy of technology, science and technology studies, and innovation policy. However, it makes unique, original contributions through its pan-African scope, emphasising African thought traditions and pragmatic arguments to enact humanistic tech futures. The book's utility spans multiple diverse audiences. It supplies essential conceptual resources and an integrative knowledge base for scholars and students. For policymakers, the principles and proposals inform extensively reforming innovation ecosystems. Finally, the accessible discussion equips ordinary citizens with conceptual tools to participate in steering technology democratically. This scope centres on constructing a holistic philosophical framework deeply rooted in African values to guide just and sustainable technological innovation across the diverse continent. By demonstrating Africa's rich knowledge heritage and applying this wisdom to tackle contemporary challenges, the book charts pathways to democratise technological futures in ethical, empowering directions.

REFERENCES

Dei, S. G. J. (2020). Elders' cultural knowledge and African indigeneity. In J. Abidogun & T. Falola (Eds.), *The Palgrave handbook of African education and indigenous knowledge*. Palgrave Macmillan. https://doi.org/10.1007/978-3-030-38277-3_14

Gwagwa, A., Kazim, E., & Hilliard, A. (2022). The role of the African value of Ubuntu in global AI inclusion discourse: A normative ethics perspective. *Patterns, 3*(4).

Hallen, B. (2002). *A short history of African philosophy*. Indiana University Press.

Hasenöhrl, U. (2021). Histories of technology and the environment in post/colonial Africa: Reflections on the field. *Histories, 1*(3), 122–144. https://doi.org/10.3390/histories1030015

Jecker, N. S., Atiure, C. A., & Ajei, M. O. (2022). The moral standing of social robots: Untapped insights from Africa. *Philosophy and Technology*, 1–22.

Kaschula, R. H., & Mostert, A. (2011). From oral literature to technauriture: What is in a name? *Unknown title*. http://www.dspace.cam.ac.uk/handle/1810/237322

Lauer, H. (2017). African philosophy and the challenge of science and technology. In A. Afolayan & T. Falola (Eds.), *The Palgrave handbook of African philosophy*. Palgrave Macmillan. https://doi.org/10.1057/978-1-137-59291-0_39

Mbiti, J. S. (1970). *African religions & philosophy*. Heinemann.

Menkiti, I. A. (1984). Person and community in African traditional thought. In R. A. Wright (Ed.), *African philosophy: An introduction* (pp. 171–181). University Press of America.

Oladumiye, B. E. (2014). Perception of relationship between art and science in contemporary African arts and technology. *International Journal of Art, Culture, Design, and Technology (IJACDT), 4*(2), 51–63.

Omoregbe, I. J. (1998). African philosophy: Yesterday and today. In E. C. Eze (Ed.), *African philosophy: An anthology*. Blackwell.

Serequeberhan, T. (1994). *The hermeneutics of African philosophy: Horizon and discourse*. Routledge.

Tutu, D. (1999). *No future without forgiveness.* Random House.
Verharen, C., Tharakan, J., Bugarin, F., Fortunak, J., Liu, M., Middendorf, G., Gutema, B., & Kadoda, G. (2014). African philosophy: A key to African innovation and development. *African Journal of Science, Technology, Innovation and Development, 6*(1), 3–12.

CHAPTER 2

Historical Origins of Technology in Africa

2.1   INTRODUCTION

This chapter chronicles Africa's technological history in a way that helps overturn the limiting assumptions we see today. Africa has a rich but overlooked history of scientific and technological innovation, from ancient astronomical systems to modern digital platforms. However, the dominant narratives coming from Europe have long dismissed the continent as primitive or backward before colonialism arrived (Adas, 1989). These stereotypes have made it harder for us to appreciate African cultures' diverse and creative technical heritage fully. This means we miss out on the skilled craftsmanship, creative ingenuity, and sophisticated knowledge systems developed in Africa over centuries. A close look at African innovation before and during colonialism shows us that advanced metallurgy, architecture, textiles, tools, medicine, and transportation networks were well-adapted to local environments and needs. This evidence counters outside depictions of Africans as lacking original technical capabilities. Documenting Africa's homegrown knowledge can help dispel myths of primitiveness and highlight the communal values underpinning technologies focused on collective advancement (Austen & Headrick, 1983). As such, revisiting Africa's rich technical legacy matters urgently today as current development challenges require innovation tailored to local contexts. Simply importing foreign technologies has often failed to meet Africa's needs. Policies frequently remain stuck in colonial notions that Western

© The Author(s), under exclusive license to Springer Nature Switzerland AG 2024
Y. Ndasauka, *African Mind, Culture, and Technology*,
https://doi.org/10.1007/978-3-031-62979-2_2

paradigms are ideologically superior. However, indigenous knowledge systems offer grassroots, sustainable solutions if we recognise and integrate them into modern policies and practices.

This chapter synthesises current research across fields to outline the African innovation timeline, including ancient accomplishments, colonial repressions, and persisting local knowledge. Extensive evidence of sophisticated pre-colonial technologies disproves primitive notions. Rather than building on African skills, imperial policies extracted labour and resources, entrenching dependence while suppressing traditional knowledge. However, adaptation, resistance, and blending of paradigms also mark this history. My goal is to reverse the erasure of African ingenuity and outlooks. This fuller context shows that today's dependence stems not from ancient backwardness but from policies that stifled capabilities (Rodney, 2018). Rediscovering Africa's diverse technical heritage and communal values is critical for transforming policies and attitudes to enable empowering, humanistic tech futures across the continent. The imperative now is to materially and cognitively decolonise technology. With its depth and reframing of assumptions, I hope this exploration of Africa's technology history can spur such paradigm shifts.

The chapter unfolds this history across three eras, offering key insights: (1) indigenous innovations developed over millennia as diverse African civilisations created technical applications, knowledge systems, and engineering skills adapted to their contexts; (2) colonial technologies were imposed to extract resources for foreign benefit, actively suppressing traditional knowledge; and (3) post-independence policies blended local and foreign tools and paradigms, with creative localisation yet ongoing dependence and uneven tech diffusion. Each section overturns limiting stereotypes and reveals nuanced foundations shaping African innovation trajectories. This account makes clear Africa's formidable but overlooked technical capabilities and communal values from which to advance technological development sustainably.

2.2   Indigenous Innovations and Technological Systems in Africa

Africa has a rich history of homegrown innovation and technology systems long before colonialism. Over centuries, diverse communities independently developed specialised techniques, engineering capabilities, and practical knowledge tailored to African environments. The evidence is extensive. Africa has a venerable tradition of metalworking, with archaeological finds showing iron smelting started 2000 years ago in Tanzania and 3000 years ago in Niger and Uganda (Twagira, 2020; Hasenöhrl, 2021). By 1000 BCE, furnace and forging techniques had spread widely across sub-Saharan Africa. Early smelting involved shaft furnaces, bellows, and pits (Alpern, 2005). Skilled African blacksmiths with honoured statuses pioneered intricate forging methods to produce ceremonial objects, functional tools, and weapons displaying sophisticated metal shaping (Thornton, 1990). They manufactured many farming, hunting, food processing, and warfare implements. Smiths also made luxury copper and bronze jewellery and artworks for elites (Thornton, 1990). West African bronze using 10% zinc alloys emerged before Europe in the ninth century CE (Gayle, 2012). Thornton (1990) noted that expanding mining from the eleventh century mass-produced more iron tools. In the early 1800s, Africa supplied an estimated one-third of the world's iron, primarily as tools (Austen & Headrick, 1983). Metalworkers' furnaces reached modern temperatures (Miller & Van Der Merwe, 1994). Swahili coast artisans used techniques like the complex Damascus steel process for decorative symbols (Alpern, 2005). The vast and sophisticated body of knowledge developed over centuries by indigenous African cultures challenges colonial-era myths and stereotypes about the people of pre-colonial Africa as primitive or backward. This indigenous knowledge demonstrates African civilisations' complex sciences, technologies, and cultural achievements before colonialism.

Equally, Africa has an impressive architectural tradition in locally adapted building arts and engineering (Wesler, 1998). Ancient Egypt and Nubia's monumental stonemasonry built majestic pyramids, obelisks, and temples using advanced engineering to position massive blocks and mathematical expertise to enable precise architectural designs (Davidson, 1994). Later kingdoms constructed fine stone palaces and public buildings with ornate arched entryways, domes, pillars, and carvings (Eluozo, 2019). In West Africa, skilled builders constructed massive earthen mosques and walls in Mali and Ghana, like the Great Mosque of Djenne, with huge mud walls and large timber beams (see Prussin, 1968; Bourgeois, 1987). The architects utilised durable, sustainable adobe bricks, compressed earth blocks, and rammed earth drawings on local soils (Munson, 1980). Thatch roofs spanned wide diameters without interior supports, and curved, domed beehive huts perfected by groups like the Zulu, enabled air circulation (Jenkins et al., 2007). Settlements like Great Zimbabwe integrated massive stone enclosures with complex drainage systems and residential areas. These harmonious, low-impact buildings were ideally adapted to local cultures, climates, and settings through architectural knowledge and techniques that produced infrastructure finely tuned to the needs of the people, environment, and location. The buildings blended seamlessly into their surroundings and supported sustainable living.

African artisans have long-standing and venerable textile craftsmanship traditions spanning the continent. Cotton spinning and weaving began along the Niger River as early as the eleventh century CE, with pioneering indigo dyeing methods developed in Africa that would eventually spread globally (Thornton, 1990). The Ashanti people of Ghana became renowned for their elaborate, vibrant kente and adinkra cloths, which conveyed spiritual concepts and social values through intricate symbols woven into the fabrics (Spring, 1989). Master dyers among the Yoruba in Nigeria created unique and decorative fabrics like adire eleko and saki using centuries-old tie-dye techniques that produced dazzling patterns (Thornton, 1990). According to Shillington (2013), West Africans employed weaving techniques using looms and wooden block printing patterns that were passed down between generations as living knowledge systems. Swahili artisans along the eastern African coast were known for their distinctive textiles featuring zigzag, geometric, and abstract motifs (Picton, 1995). From the exquisitely embroidered cotton robes of North Africa to the prestige strips of cloth woven by the people of Central Africa, the diversity of textile designs, materials, and production methods demonstrated remarkable ingenuity, skill, and artistry developed across the continent over many centuries, dispelling colonial-era stereotypes of African cultural backwardness.

In largely agrarian societies across pre-colonial Africa, smallholder farmers and communities developed sophisticated tools and techniques tailored to local environmental conditions in order to sustain productive agriculture (McDougall, 1990). Iron hoes, axes, and sickles enabled effective land cultivation. At the same time, practices like manure application, planting patterns, terracing, and irrigation served to preserve and enhance soils and maximise water retention (Eluozo, 2019). The finger planter and other multi-pronged digging forks designed for African soils in Zimbabwe allowed precise seed placement. These raised productivity when combined with the ready availability of iron hand tools (Randall-MacIver, 1906).

Unique irrigation methods like Tanzania's sagai wheel and Nigeria's ingenious subsurface canal systems challenged assumptions about lacking agricultural technology in Africa (The Confucian Weekly Bulletin, 2020). Herding groups in semi-arid areas perfected sustainable land use practices, while agroforestry provided essential nutrition and income for many communities (Wikipedia, 2023). These diverse innovations and adaptations fuelled prosperous food production systems across pre-colonial Africa. Developing complex calendars to guide agricultural activities and manage resources further disproves stereotypical notions of African primitiveness. The ingenuity and scientific knowledge underpinning these agricultural advances reflect African civilisations' sophistication prior to colonialism.

Indigenous African medical knowledge developed over millennia, focusing on plant-based treatments, bone-setting, midwifery, and divination practices for diagnosis and healing (Harley, 1941). Ancient Egyptian doctors were pioneers in areas like diagnosis, anatomy, and pharmaceutical recipes as early as 5000 years ago (Strouhal, 1989). In medieval Europe, West African medicinal soap was highly prized for its efficacy (Kananoja, 2021). Detailed Nigerian plant pharmacopoeias document generations of curative treatments, while Tuareg, Akan, Khoikhoi, and other healers dispensed effective mineral, clay and botanical solutions that predated modern medicine (Watkins, 2021). Skilled midwives facilitated childbirth across Africa for thousands of years (Strouhal, 1989). Treatments like the Xhosa's umhlonyane and Zulu's 'cancer bush' demonstrate the extensive pharmacological expertise developed collectively by African cultures to promote communal well-being (Flint, 2008). This body of knowledge counters colonial-era tropes about primitive African healthcare. Thoroughly recording these traditional plant medicines provides a critical foundation for socialised and holistic medicine today. From bone-setting techniques to malaria remedies, the empirical observations and plant science expertise underpinning indigenous African healing practices reflect a sophisticated understanding of human biology and the medicinal properties of local flora—a vital epistemological legacy predating the arrival of Europeans by millennia.

Well before the emergence of European colonialism, African civilisations had already developed remarkable large-scale transportation infrastructure and trade networks that facilitated mobility, commerce, and communication over immense distances. In ancient Egypt, sophisticated river vessels traversed the great Nile waterway, while seaworthy ships navigated the Mediterranean carrying goods and people between the Egyptian empire and its trading partners. Impressive trans-Saharan caravan routes enabled camels to trek vast distances across the expansive desert, connecting North Africa to sub-Saharan kingdoms and cities. Wheeled transportation using carts and wagons drawn by horses and oxen also flourished in certain African regions (see Köpp-Junk, 2016; Law, 1980). Specialised dhow ships efficiently shuttled along the eastern coast between major Swahili port cities on the Indian Ocean, laden with cargo (Mabogunje et al., 2023). Overland transportation relied on well-established pathways carved through the terrain, allowing transport by cart and foot. In the dense forests of West Africa, highly adapted streamlined canoes provided passage through the myriad lakes, rivers, and wetlands that characterised the inland areas (Mabogunje et al., 2023). Porterage services developed in some regions, where hired porters transported goods between different ecological zones from coast to forest to savanna (Köpp-Junk, 2016). Large-scale water relay systems also passed resources over considerable areas (Utsua, 2015). This infrastructure enabled the vigorous trade, connectivity, and mobility essential for African civilisation.

Sophisticated communication systems enabled the transmission of messages across vast distances with speed and accuracy in pre-colonial Africa. The iconic talking drums of central Africa and Yoruba dundun drums incorporated subtle acoustic qualities that could mimetically represent the tonal inflexions of African languages. This allowed drummers to rapidly convey complex and nuanced information over kilometres, functioning as an early form of telecommunication (Lewin-Richter, 1958). Animal horns, whistles, and other instruments also encoded communicative meanings into their sounds. Additionally, specialised trade languages were developed that allowed commerce and cultural exchange between diverse linguistic groups inhabiting a region (Utsua, 2015). West Africa's sprawling trade networks fostered lingua francas like Dyula and Hausa, while Swahili connected ethnicities along 2000 miles of East African coastline. These innovative transportation and communication systems facilitated the vigorous trade, interconnection, and mobility essential for the flourishing of pre-colonial African civilisations, enabling goods, people, ideas, and innovations to circulate across the continent. From the solitary drum signalling a royal announcement to intricate symbolic meaning encoded in rhetoric, the ingenuity underpinning indigenous communication methods in Africa reveals technological and linguistic sophistication predating the arrival of Europeans by centuries.

The examples explored here offer glimpses into the immense diversity and ingenuity of knowledge systems, technologies, and infrastructure independently developed and operated across African cultures before colonisation. These innovations reveal how local needs shaped adaptive grassroots solutions well-aligned with communal values and harmonious co-existence with the natural environment. Of course, broad generalisations cannot fully encapsulate the nuances and depths of African creativity across spheres. However, the cases presented underscore how depictions of supposed "primitive" conditions fundamentally mischaracterise and ignore the many indigenous advancements present in Africa before European arrival. Contrary to colonial-era stereotypes, pre-colonial African civilisations possessed remarkably complex knowledge systems, technical skills, and sophisticated networks on a par with other ancient societies worldwide, yet uniquely configured to sustain human life and foster communal thriving amid the continent's myriad landscapes, climates, and resources. This monumental legacy remains a vital wellspring of inspiration, identity, and strength for Africans today, with traces of ancient ingenuity surfacing in modern local innovation and adaptation. Comprehending the richness of Africa's indigenous heritage holds relevance as African nations seek more just, equitable, and self-directed futures grounded in the revival and leverage of their formidable cultural strengths and epistemologies.

2.3   Technology Under Colonialism in Africa

The colonial era ushered in fundamental changes in how technology was imposed across Africa in ways that disempowered local communities. External paradigms and tools were forced on the continent mainly to serve imperial resource extraction and exploitation, not to meet indigenous needs holistically. A core colonial idea was that Europeans were culturally and technologically superior to 'backward' Africans. Colonisers branded generations of indigenous knowledge designed to solve local challenges as ignorant or mere 'witchcraft'. Europeans made barely any effort to understand the logic, validity, or communal value underpinning these localised systems developed in Africa over centuries. Colonisers denied that Africans were capable of original innovation, portraying the continent as plagued by ignorance and stagnation before the supposed 'civilising mission' of colonialism arrived. This self-serving narrative justified attacking and dismantling the foundations of pre-colonial African knowledge systems to implant foreign paradigms.

European colonisers deliberately limited educational and technological opportunities for Africans as a means to preserve racial inequalities and claims of superiority. Under colonial rule, technical training for Africans focused on basic vocational skills like agriculture, mechanics, surveying, and clerical work to fill lower-level positions. Additional training was contingent on Africans adopting Western culture and religion rather than building on indigenous knowledge systems (Emeagwali, 2003). Under colonial policies, secondary and higher education enrolment was kept minimal across the continent. Missionaries sought to replace indigenous cosmologies, histories, sciences, arts, and educational practices wholesale with European Christian curricula, further devaluing Africa's rich communal learning traditions (Emeagwali, 2003). Schools aimed to provide only rudimentary technical skills to serve low-level colonial economic jobs, not to empower local expertise or social mobility. Literacy programmes rigidly emphasised Bible study rather than works of literature, science or technology, obstructing African languages' ability to convey complex technical ideas (Emeagwali, 2016). This gross underinvestment in holistic African education and technical training under colonialism further entrenched dependence on the colonisers. It constrained opportunities for Africans to gain higher-level skills that would enable technological self-sufficiency, economic competitiveness, or social empowerment. The few who accessed education faced pressures to adopt Western paradigms and reject indigenous worldviews. These policies delayed African technological and industrial advancement for generations by depriving its people of education suited to their context, needs, and potential.

Indigenous agricultural techniques refined over centuries were dismissed as 'unscientific' or wasteful by colonisers despite their limited knowledge of tropical farming (Adas, 1989). Restrictive colonial forest policies disrupted community access to ancestral lands, undermining sustainable gathering and grazing practices. Smallholder farmers' wisdom was branded superstitious and not recognised as grassroots innovation enabling self-reliance (Adas, 1989). Colonial officials proposed replacing communal land rights with privatisation to 'modernise' farming, even though traditional shifting cultivation had prevented erosion and pests for generations (Katzung, 2020). Colonial authorities also prohibited Africans from fully participating in major export crops or industries to avoid economic competition. For example, in Kenya, native coffee farming was banned to protect white settler plantations (Parker, 1952). Racially differentiated land policies barred Africans from freehold rights or advanced agricultural technologies. Technologies like tractors, irrigation, and processing machinery were restricted to white farms, enabling their commodity export domination (Anderson & Throup, 1985). Displaced African farmers were relegated to marginal lands without productivity-enhancing technologies (Parker, 1952).

Colonial authorities outlawed traditional African methods and crops in favour of mono-crop plantations of commodities like cocoa, coffee, cotton, and palm oil grown for export markets in Europe. This enforced shift towards plantation-style commercial farming focused on exporting single cash crops exhausted soils and diverted labour away from producing traditional food staples for local populations, causing widespread famines across colonised Africa (Roessler et al., 2022). Indigenous practices like mixed cropping, fallowing of fields, and integration of nitrogen-fixing trees preserved soil fertility over generations, while communal tenure systems aligned cultivation with seasonal rhythms and needs. However, colonial administrations often replaced these localised land management strategies with extractive policies to maximise agricultural commodity exports. In these myriad ways, disruptive colonial agricultural policies, technologies, and infrastructures undermined indigenous farming knowledge honed in Africa over centuries, causing lasting environmental and social damage.

Indigenous African communities had developed sophisticated systems of sustainably managing forest resources through agroforestry, foraging, hunting, and low-impact timber harvesting over generations. But these localised approaches were often disrupted under colonial rule. Colonial administrations prioritised the rapid extraction of forest products like valuable hardwood timbers, rubber, and minerals for export to Europe over local communities' sustainable use of resources. Colonial governments approved extraction-focused commercial logging concessions and mining operations without regard for indigenous use rights or environmental impact (Neumann, 2002). Resource management for community benefit was replaced by exploitation for imperial commercial gain. Timber companies cleared old-growth forests that local people depended on for hunting game, gathering medicinal plants, and producing food crops. Sacred groves held by villages for centuries were razed. Uncontrolled logging caused extensive erosion and disrupted traditional agroforestry systems integrating trees with agriculture. In these ways, indigenous knowledge of local ecosystems enabled sustainable management of forests through hunting, foraging, agroforestry, and low-impact harvesting. However, colonial powers disrupted these communal systems by imposing extractive technologies prioritising commercial exploitation over environmental protection or community needs.

Before European colonialism, many African societies had developed decentralised, participatory governance structures and cooperative economic models that enabled inclusive grassroots advancement for centuries. But these indigenous systems were actively dismantled under colonial rule in ways that disempowered local communities. Colonial powers imposed authoritarian rule and bureaucratic administration that concentrated decision-making power in the hands of imperial governors, dismissing pre-existing African political institutions as 'primitive' (Rodney, 2018). Cooperative economic structures grounded in communal values were replaced by exploitative colonial economies oriented around resource extraction for European profit. Dismantling local governance and shared economic models that had coordinated activities like farming, herding, and craft production for generations allowed colonial regimes to increase control over indigenous labour and resources. Collective ownership of land and resources by villages was replaced with private property rights favouring colonists. Chief's authority was replaced by authoritarian appointees loyal to colonial administrations. European courts and punitive justice superseded customary legal systems for conflict mediation and restorative justice. Loss of communal lands severed ties to ecological wisdom developed over centuries. This decline in cooperative structures greatly undermined knowledge-sharing and long-held governance systems.

Cheap imported European textiles undermined African economies and local artisanal production, which undersold indigenous craftspeople and flooded markets (Rodney, 2018). While colonialism relied on African mining, farming and labour, resource extraction and infrastructure technologies served imperial interests. Railways, ports, and roads aimed to export raw materials to Europe efficiently, not to support local development (Headrick, 1979). Agricultural policies mandated cash-cropping for export rather than diverse food production for African needs (Roessler et al., 2022). Grid electrification powered colonial plantations, mines, and administration but left rural Africans unconnected (Headrick, 1979). Manufacturing industries were discouraged to prevent competition with Europe, while imported goods stifled African industrial growth (Rodney,

2018). Segregated urban planning privileged amenities and services in European zones while neglecting African neighbourhoods (Njoh, 2009). These infrastructural decisions reflect how technology advanced colonial profit over inclusive African development.

Before colonialism, indigenous medicine in Africa had developed a sophisticated understanding of local pharmacology and holistic healing arts over generations. But traditional practitioners faced prohibitions under colonial rule, resulting in an immense loss of plant pharmaceutical knowledge and community healthcare capacity. Colonial authorities banned traditional healers and herbalists, declaring Western biomedicine the only approved orthodoxy. This served to delegitimise indigenous expertise developed over centuries, even when some diseases were exacerbated by colonialism itself (Turshen, 1977). While biomedicine advanced sanitation and treated some conditions, its infrastructure primarily served colonial interests. Health interventions like clinics and hospitals largely benefited colonial administrators, military forces and local workforces necessary to sustain imperial operations rather than universally serving African needs (Turshen, 1977). Dismantling indigenous medical knowledge disrupted essential community care, especially for women and rural villagers. Dependence on expensive imported medicines was created. Loss of plant expertise has long-term consequences, as this empirical knowledge provided low-cost medicinal solutions synthesised with a deep understanding of local ecologies over generations. Some colonial doctors even appropriated herbal knowledge without acknowledgement, further erasing African contributions to healing arts.

Colonial regimes imposed bureaucratic obstacles that obstructed African entrepreneurship and technology skills acquisition. Exclusionary policies ensured ongoing dependence on European expertise long after political independence. Additionally, racially preferential lending policies hampered African business expansion and technological upgrading. Furthermore, the substantial racial wage gap made capital accumulation extremely difficult for Africans to achieve under colonialism. As a result, these interlocking obstacles to African participation in higher-level technical roles enabled colonisers to monopolise economic privileges and perpetuate myths of African inadequacy over generations. Consequently, they created enduring disparities in technological capabilities between the colonised and coloniser that still linger today (Fanon, 1961). Moreover, the systematic suppression of indigenous knowledge forms, constraints on adopting new technologies curtailed African economic opportunities, and obstruction of tertiary education severely limited possibilities for African technological advancement under colonial rule. Specifically, technical skills development was narrowly confined to routine manual tasks serving colonial production needs rather than empowering local creativity or ingenuity. Thus, the result was enduring dependence on external Western expertise, imported machinery from Europe, and primary commodity export models largely dictated by colonial interests. These policies severely hampered Africa's potential for independent industrialisation and innovation. These historic capability constraints helped shape lasting developmental challenges across the continent. Moving forward, reclaiming technological sovereignty requires uplifting marginalised African knowledge forms and undoing the lingering impacts of colonial obstruction.

For centuries, indigenous technologies supporting self-sufficient African communities were often displaced or disrupted under colonial rule. In the name of 'modernisation', locally adapted indigenous technologies were dismissed as backwards and inferior. Unfortunately, these ambivalent colonial attitudes towards African technology persist problematically in modern development policies. Under colonialism, technology primarily enabled the exploitative extraction of African resources and labour to benefit foreign powers. Technology served production but not skills development for Africans. Infrastructure facilitated the appropriation of the most productive lands by settlers. Colonisers utilised technology for extraction, control, and domination rather than African empowerment.

2.4   Technology in Post-independence Africa

When African nations gained independence in the 1950s–1970s, they faced significant decisions about technology and development. Alongside rapid modernisation aims, countries also grappled with promoting indigenous identity and self-reliance after colonialism suppressed traditional knowledge. Newly free states prioritised modern technology for nation-building and economic growth. Huge investments went into infrastructure like roads, telecoms, factories, dams, and social services to benefit citizens and catalyse development. Technology transfer from abroad seemed the best solution for quickly building human and industrial capacity at scale. Borrowing from China, African governments created public companies to manage critical sectors like mining, energy, and transportation (Asingia, 2019). We saw experts assist in rapid skills transfer until locals were trained and multinationals equipped factories while foreign aid financed imported infrastructure. Governments actively facilitated technology inflow for rapid results, prioritising speed over self-reliant capacity building.

In addition, African governments and entrepreneurs creatively localised many imported technologies for indigenous development to fit local needs, resources, and environments (Asingia, 2019). For instance, smallholder farmers integrated new tools like pumps and fertilisers without displacing communal practices like gravity irrigation and manure. In medicine, scientists combined traditional herbal medicine with modern pharmaceuticals to improve community health access. These adaptations aimed to make foreign techniques more accessible in Africa, like the Kenya Ceramic Jiko stove's use of local clays for efficient fuel use. This creative localisation shifted from mere imitation towards innovation in African terms. However, dependence on imported inputs and designs abroad still constrained advancement. Further, African governments relied on external systems designed for Western contexts. These tendencies seemed to perpetuate dependence on foreign companies and advisors. This resulted in most projects failing to build local innovation skills. Though intended to accelerate progress, uncritical technology transfer often proved inappropriate and unsustainable. Consequently, adopting and adapting technology for comprehensive advancement remained an ongoing challenge.

To counter inappropriate external dependency, African scholars advocated reviving indigenous innovation and philosophies as foundations for autonomous development (Forje, 1987). For example, African scholars proposed revitalising indigenous water harvesting to address shortages. Further validating marginalised herbal medicine, ecological farming, and communal economics would reform policies to recognise their contemporary value. An example is South Africa's 2004 Indigenous Knowledge Systems Policy, which promotes integrating such endogenous and modern knowledge in R&D, education, and business. Regional networks like the African Technology Policy Studies Network also research leveraging traditional knowledge for sustainable development. Blending indigenous and appropriate modern technologies would enable context-suitable innovation. However, these initiatives were marred by deep external reliance, shaping uneven technology diffusion across the continent.

The colonial experience significantly disrupted indigenous innovation systems across Africa. Before colonial rule, many societies had rich artistic, agricultural, and engineering traditions uniquely adapted to local contexts. However, colonial policies often denigrated and suppressed these knowledge forms in favour of Western science and technology. This legacy continued to hinder the recognition and development of African ingenuity after independence. Additionally, cultural attitudes from the colonial era have hindered recognition of African ingenuity. Racist assumptions about the inferiority of indigenous knowledge led many to dismiss local technologies and abilities. These prejudices carried over even after independence, influencing education policies and national narratives. As a result, policies emphasising foreign technology transfer often overlook building local capabilities. Despite rhetoric about independence and Africanization, western paradigms remain prized over traditional knowledge forms in many sectors. In national curricula, for instance, science and technology subjects prioritise indigenous technologies and histories more than Western science. So, while political leaders speak about cultural relevance, Western norms continue to dominate fields like engineering, medicine, and agriculture.

2.5   Unfulfilled Dreams: Technology Blending African and Western Worldviews

While African nations adopted many foreign technologies after colonialism, efforts emerged to integrate external tools with indigenous knowledge. At their best, Africanized technologies synthesise modern scientific techniques with cultural philosophies and grassroots social practices to enable inclusive, sustainable development. In construction, for example, some builders have creatively adapted techniques from abroad using local materials and site planning wisdom to build infrastructure attuned to community needs. In Nigeria, a local engineer pioneered modular housing designs using local clays, natural cooling, and shared space planning from villages to create affordable, culturally resonant modern buildings (Okpala, 1983). Some architects integrated fractal shapes from African textiles and nature with contemporary materials for buildings reflecting cultural identities. These examples demonstrate the power of blending vital elements of traditional knowledge and appropriate external technologies to address community needs holistically. Similarly, in agriculture, equipment-sharing cooperatives and traditional crop markets sustain productivity and communities instead of rigid technocratic prescriptions. Although often a failed policy, decentralised renewable energy models have also shown potential to expand electricity access in rural areas. By tapping into indigenous self-help traditions, community-led microgrids could provide locally managed power, enabling income generation and empowerment.

Despite this, the lingering dismissal of African worldviews as 'primitive' hinders fully integrating their innovation principles into formal institutions and education. Reluctance to validate non-Western knowledge limits recognising indigenous farmers as experts. Additionally, the lack of policies around communal energy management constrains scaling grassroots solutions. Attempts are being made at culturally attuned digital inclusion to marry oral traditions, peer learning and collective governance with new media tools. Rural digital hubs foster participatory local content creation. Integrating appropriate tech with African lifestyles is possible but hampered by institutional inaction. Visions of Africanized technology empower communities as agents, not just consumers. Further, relegating African knowledge to 'culture' rather than 'science' hinders its integration into formal institutions guiding technology use and education. As I will discuss, Ubuntu thinking, which prioritises collective advancement over individual gain as technology's purpose, may be utilised to shape technology for human development.

Colonial legacies of devaluing Africa's intellectual heritage persist, limiting the continent's technological autonomy and agency. Genuinely fulfilling the creative promise of Africanizing technology ultimately requires moving beyond blending tools to transforming mindsets. It necessitates embracing African philosophies emphasising communal advancement and harmony with nature as reforming principles. This demands the decolonisation of formal institutions to legitimise indigenous worldviews as valid technological knowledge, not just cultural artefacts. Africa's rich wellsprings of bio-cultural diversity must be recognised as assets for contextual innovation. Though difficult, centring African philosophies and knowledge traditions holds transformational potential for technology to serve inclusive, sustainable human development.

Decades after formal decolonisation, Africa's economic landscape remains dominated by primary commodity exports and imported technology designed externally. Critical sectors like Africa's telecoms, machinery, chemicals, and transport rely on foreign tech for 80%–90% of supply. We only recently saw one of the first Africa Mobile Phone factories in Kenya. This dependence constrains advancing endogenous scientific capabilities, technical skills, and manufacturing. Progress in local production and innovation has been minimal. Pervasive tech reliance, from infrastructure to consumer goods, reflects institutional legacies of colonialism that still permeate the present. Colonial policies stunted indigenous industry and channelled colonies to provide raw materials and import European manufactured goods. This extractive pattern continued after independence, as African nations exported more commodities but imported technologies abroad. In the decades since short-term output expansion masked stagnation in productivity and self-driven advancement. As argued previously, commodity exports brought revenues but little technical learning.

Dependence on external technology has many costs for Africa. Foreign donors heavily influence technology priorities in ways that frequently misalign projects with actual community needs. For example, Angola's oil wealth often fuelled high-end consumption more than balanced human development priorities for its people (Hammond, 2011). More broadly, African nations remain trapped in economic structures optimised for colonial-era extraction and export rather than local capability building. This makes structurally transforming their technological capabilities on their own terms challenging. Unevenness persists as cities boast pockets of innovation while rural areas lag behind. The sobering truth is that South Africa and Kenya lead in digital entrepreneurship, but regional and national disparities across the continent are stark. Integration into global tech production still mirrors unequal colonial geographic patterns rather than enabling self-directed advancement. This is evidence that accessing cutting-edge technologies alone does not automatically confer capacities for sustainable progress.

2.6   CONCLUSION

This chapter has explored Africa's overlooked history of science and technology innovation in hopes of overturning dominant stereotypes of precolonial primitiveness and postcolonial backwardness. Contrary to dismissive assumptions, diverse African civilisations independently developed creative innovations, engineering systems, and knowledge infrastructures that enabled thriving cultures before European contact. Sophisticated metallurgy, architecture, manufacturing, medicine, agriculture, and trade networks grew from communal values using local resources to meet needs. This formidable legacy counters external depictions of Africans lacking technological capabilities before colonialism. However, the violent imposition of foreign paradigms under colonialism actively suppressed indigenous knowledge through repression and devaluation. External technologies were implemented primarily for imperial control and resource extraction, not for developing self-reliant African capacities. Actively obstructing traditional practices and education entrenched dependence by systematically stifling capabilities. Colonial ambivalence towards fostering African skills cast a long shadow after independence. Post-independence policies blended some exogenous technologies with efforts to revive and revalue endogenous innovation and philosophies. However, entrenched institutional legacies obstruct fully transitioning from imitation to creative localisation and self-driven development. Unfinished liberation from dependent accumulation patterns inherited from colonialism sustains uneven diffusion of technologies essentially designed abroad. Ongoing reliance on outside expertise inhibits endogenous advancement.

This fuller account shows that Africa's current development challenges do not originate from ancient backwardness but from extractive colonial policies and persisting Eurocentric biases that still devalue African knowledge today. Recognising Africa's technical achievements and communal philosophies is vital to overturning such limiting cognitive frameworks. The diverse knowledge systems developed over centuries to serve African societies, not exploit them, hold overlooked value for solving contemporary problems. Transforming these assumptions is essential for such cognitive and epistemic liberation. Africa's history reveals remarkable ingenuity, complex knowledge, and communal ethics. The continent's development challenges stem not from a lack of innovation potential but from histories and ideologies that suppressed capabilities. Africa's diverse creative communities provide wellsprings for envisioning technologies focused on collective advancement. This liberating recognition of Africa's achievements and outlooks is indispensable for building technological self-determination. The nuanced history recounted here makes Africa's formidable homegrown capabilities undeniably evident.

REFERENCES

Adas, M. (1989). *Machines as the measure of men*. Cornell University Press.

Alpern, S. B. (2005). Did they or didn't they invent it? Iron in Sub-Saharan Africa. *History in Africa, 32*, 41–94. http://www.jstor.org/stable/20065735

Anderson, D., & Throup, D. (1985). Africans and agricultural production in colonial Kenya: The myth of the war as a watershed. *The Journal of African History, 26*(4), 327–345. http://www.jstor.org/stable/181653

Asingia, B. (2019). *The last digital frontier: The history and future of science and technology in Africa.* Blackwells Publishing.

Austen, R. A., & Headrick, D. (1983). The role of technology in the African past. *African Studies Review, 26*(3/4), 163–184. https://doi.org/10.2307/524168

Bourgeois, J. (1987). The history of the great mosques of Djenné. *African Arts*, UCLA James S. Coleman African Studies Center, *20*(3), 54–92. https://doi.org/10.2307/3336477. https://www.jstor.org/stable/3336477.

Davidson, B. (1994). *The search for Africa: History, culture, politics* (p. 57(8)). Random House. ISBN 0-8129-2278-6.

Eluozo, C. (2019). Science and technology in Africa: A historical perspective. *International Journal of Innovative Social & Science Education Research, 7*(1), 80–87.

Emeagwali, G. (2003). African indigenous knowledge systems (AIK): Implications for the curriculum. In T. Falola (Ed.), *Ghana in Africa and the world: Essays in honour of AduBoahen*. Africa World Press.

Emeagwali, G. (2016). Intersections between Africa's indigenous knowledge systems and history. In G. Emeagwali & G. Sefa Dei (Eds.), *African indigenous knowledge and the disciplines* (pp. 1–17). Sense Publishers.

Fanon, F. (2004) [1961]. *The Wretched of the Earth.* Translated by Philcox, Richard. With a foreword by Bhabha, Homi K. Grove Press.

Flint, K. E. (2008). *Healing traditions: African medicine, cultural exchange, and competition in South Africa, 1820-1948.* Ohio University Press. ISBN 9780821418499.

Forje, J. W. (1987). Technological change and the development of Africa. *Présence Africaine, 143*, 122–140. http://www.jstor.org/stable/24351530

Gayle, D. (2012). *How ancient Africans were the first nerds: Birth of technology traced back 70,000 years to the continent's southern tip.* The Daily Mail. Accessed through https://www.dailymail.co.uk/sciencetech/article-2243946/How-ancient-Africans-nerds-Birth-technology-traced-70-000-years-continents-southern-tip.html

Hammond, J. L. (2011). The resource curse and oil revenues in Angola and Venezuela. *Science & Society, 75*(3), 348–378. http://www.jstor.org/stable/41290174

Harley, G. (1941). *Native African medicine with special reference to its practice in the Mano tribe of Liberia* (p. 26). Harvard University Press. ISBN 978-0-674-18304-9. OCLC 598805544.

Hasenöhrl, U. (2021). Histories of technology and the environment in post/colonial Africa: Reflections on the field. *Histories, 1*(3), 122–144. https://doi.org/10.3390/histories1030015

Headrick, D. R. (1979). The tools of imperialism: Technology and the expansion of European colonial empires in the nineteenth century. *The Journal of Modern History, 51*(2), 231–263. http://www.jstor.org/stable/1879216

Jenkins, T., Bonner, P., & Esterhuysen, A. (2007). *A search for origins: Science, history and South Africa's 'Cradle of Humankind'*. NYU Press. ISBN 9781776142309.

Kananoja, K. (2021). *Healing knowledge in Atlantic Africa*. Cambridge University Press. ISBN 978-1-108-49125-9.

Katzung, H. B. R. (2020). *Oppression and dispossession out of fields of plenty: Colonialism and indigenous agricultural transformation* (p. 824). Student Publications.

Köpp-Junk, H. (2016). Wagons and carts and their significance in ancient Egypt. *Journal of Ancient Egyptian Interconnections, 9*, 14.

Law, R. (1980). Wheeled transport in pre-colonial West Africa. *Africa: Journal of the International African Institute, 50*(3), 249–262. https://doi.org/10.2307/1159117

Lewin-Richter, M. E. (1958). *The Akan of Ghana: Their ancient beliefs*. Faber & Faber.

Mabogunje, A. L., Nicol, D. S. H. W., Middleton, J. F.M., Steel, R. W., Kröner, A., Gardiner, R. K. A., Dickson, K. B., Clarke, J. I., Smedley, A., & McMaster, D. N. (2023, October 17). *Africa*. Encyclopedia Britannica. https://www.britannica.com/place/Africa

McDougall, E. A. (1990). Production in precolonial Africa. *African Economic History, 19*, 37–43. https://doi.org/10.2307/3601890

Miller, D. E., & Van Der Merwe, N. J. (1994). Early metal working in Sub-Saharan Africa: A review of recent research. *The Journal of African History, 35*(1), 1–36. http://www.jstor.org/stable/182719

Munson, P. J. (1980). Archaeology and the prehistoric origins of the Ghana Empire. *The Journal of African History., 21*(4), 457. https://doi.org/10.1017/s0021853700018685

Neumann, R. P. (2002). *Imposing wilderness: Struggles over livelihood and nature preservation in Africa*. University of California Press Books.

Njoh, A. J. (2009). Urban planning as a tool of power and social control in colonial Africa. *Planning Perspectives, 24*(3), 301–317. https://doi.org/10.1080/02665430902933960

Okpala, D. C. I. (1983). Promoting the use of local building materials in Nigerian house construction: Problems and prospects. *Ekistics, 50*(298), 42–46. http://www.jstor.org/stable/43620610

Parker, G. G. (1952). British policy and native agriculture in Kenya and Uganda. *Agricultural History, 26*(4), 125–131. http://www.jstor.org/stable/3740472

Picton, J. (1995). *The art of African textiles: Technology, tradition, and lurex*. Becker, Rayda., Barbican Art Gallery. ISBN 0853316821. OCLC 34052769.

Prussin, L. (1968). The Architecture of Islam in West Africa. *African Arts*, UCLA James S. Coleman African Studies Center, *1*(2), 32–74. https://doi.org/10.2307/3334324. JSTOR 3334324.

Randall-MacIver, D. (1906). *Mediæval Rhodesia*. Macmillan and Co..

Rodney, W. (2018). *How Europe underdeveloped Africa*. Verso.

Roessler, P., Pengl, Y. I., Marty, R., Titlow, K. S., & Van de Walle, N. (2022). The cash crop revolution, colonialism and economic reorganization in Africa. *World Development, 158*, 105934. https://doi.org/10.1016/j.worlddev.2022.105934

Shillington, K. (2013). *Encyclopedia of African History* (Vol. 3, pp. 736–737). Routledge. ISBN 978-1-135-45670-2.

Spring, C. (1989). *African textiles*. Crescent.

Strouhal, E. (1989). *Life in ancient Egypt* (p. 243). University of Oklahoma Press. ISBN 0-8061-2475-X.

The Confucian Weekly Bulletin. (2020). *Technological Developments in Pre-Colonial Africa*. Accessed through https://confucianweeklybulletin.wordpress.com/2020/02/17/technological-developments-in-pre-colonial-africa/

Thornton, J. (1990). Precolonial African industry and the Atlantic Trade, 1500-1800. *African Economic History, 19*, 1–19. https://doi.org/10.2307/3601886

Turshen, M. (1977). The impact of colonialism on health and health services in Tanzania. *International Journal of Health Services, 7*(1), 7–35. http://www.jstor.org/stable/45129974

Twagira, L. A. (2020). Introduction: Africanizing the history of technology. *Technology and Culture, 61*(2), S1–S19. https://doi.org/10.1353/tech.2020.0068

Utsua, T. P. (2015). The evolution of African indigenous science and technology. *International Knowledge Sharing Platform, 16*.

Watkins, C. (2021). *Palm Oil Diaspora: Afro-Brazilian landscapes and economies on Bahia's Dendé Coast*. Cambridge University Press. ISBN 978-1-108-80829-3. [page needed].

Wesler, K. W. (1998). *Historical archaeology in Nigeria* (p. 143, 144). Africa World Press. ISBN 0-86543-610-X, 9780865436107.

Wikipedia. (2023). *History of science and technology in Africa*. In Wikipedia. https://en.wikipedia.org/wiki/History_of_science_and_technology_in_Africa

CHAPTER 3

Decoding the Digital Revolution

3.1 Introduction

This chapter analyses the socio-technical factors influencing the adoption and integration of technologies in Africa. It examines how digital platforms, AI, and other innovations spread unevenly, creating digital divides and development gaps between groups based on gender, geography, age, and class. Moreover, the chapter applies critical perspectives like Afrocentric and postcolonial theories to highlight how external technical models often fail to align with African realities. It discusses cases showing the intended benefits of technologies accompanied by unintended negative social consequences. Consequently, the chapter argues for reforming innovation ecosystems to foster ethical assessment and adaptation of technologies to maximise social good in African contexts. It calls for user-centred design approaches that meet local community needs, values, and environments. Furthermore, strategies for the participatory development of appropriate, sustainable technologies are recommended to increase social embedding. The chapter demonstrates how ethical evaluation, co-creation, and reconfiguration of foreign innovations can accelerate inclusive tech diffusion. Ultimately, technology adoption can be transformed from the imposition of inconsistent systems to actively integrating appropriate, empowering solutions designed for African advancement.

Technology adoption and integration in Africa have seen transformational successes; however, uneven diffusion patterns persist, with

Y. Ndasauka, *African Mind, Culture, and Technology*,
https://doi.org/10.1007/978-3-031-62979-2_3

significant access gaps across countries and demographic groups. This complex landscape of technology outcomes reflects the interplay between multifaceted drivers, barriers, and impacts shaped by Africa's diverse locales. Thus, analysing these dynamics provides insights into optimising technology's contributions to sustainable development in the region. Technology uptake and spread in Africa have strong linkages with the availability of community resources and capabilities, infrastructure accessibility, user attitudes and priorities, demographic differences, and contextual relevance. For instance, education levels, digital skills, electricity access, and cultural attitudes are recurring factors influencing adoption outcomes. Additionally, patterns of uneven diffusion emerge around income, gender, urban-rural, and regional divides. At the same time, technologies demonstrate tangible benefits for financial inclusion, entrepreneurship, service delivery, and productivity where effectively integrated. However, concerns remain around growing inequality, dependence on foreign solutions, cultural disruption, and environmental externalities.

This multidimensional terrain underscores that technology adoption pathways in Africa cannot be understood through simple diffusion models. Instead, integrative analytical approaches recognising community-specific enablers and barriers across technical, infrastructural, social, economic, and cultural dimensions are imperative. Most studies concur that achieving optimal, responsible, and inclusive technology integration in Africa requires a grounded understanding of user priorities through participatory processes, localisation, capacity building, and access models tailored for affordability and sustainability challenges. Firstly, the chapter assesses technology outcomes in Africa across adoption determinants, diffusion patterns, unintended consequences, and constraints. Synthesising these insights, the chapter first surveys key arguments on how community resources, infrastructure access, costs, user attitudes, and demographics shape uneven technology uptake. Subsequently, it analyses regional diffusion variations and adoption gaps across digital divides. Next, it discusses the developmental impacts and risks of emerging technologies. Finally, the chapter examines critical constraints to appropriate technology adoption and potential inclusive innovation strategies.

3.2   FACTORS INFLUENCING TECHNOLOGY ADOPTION

Multidimensional factors, including available resources, infrastructure accessibility, affordability considerations, user perceptions and attitudes, relevance to local contexts, and demographic differences, shape technology adoption across Africa. A complex interplay of these social, economic, cultural and structural enablers and barriers underlies uneven technology diffusion patterns across the continent. Therefore, responsible innovation requires understanding key determinants mediating adoption to guide appropriate, sustainable, inclusive technology integration (Amankwah-Amoah, 2016; Asongu & Odhiambo, 2019).

A community's existing resources and capabilities shape technology adoption prospects. Specifically, technical skills, digital literacy, educational levels, and complementary infrastructure must reach sufficient thresholds to enable the utilisation of new systems. Moreover, managerial and maintenance expertise further support integration. In fact, many technologies require substantial human capabilities before implementation. For instance, limited education constrained widespread computer adoption in early Internet diffusion across Africa, though skill building enabled later engagement. Consequently, uneven digital skills sustain tech usage divides. However, contextual relevance and usability can sometimes compensate for skills gaps, as with widespread mobile phone adoption leveraging accessible voice interfaces (Aker & Mbiti, 2010).

Insufficient financial resources also hinder technology acquisition by individuals, firms, and governments. Notably, capital costs of technologies like smartphones or industrial machinery restrain adoption absent financing mechanisms. Microfinancing and sharing models like equipment cooperatives lower barriers through incremental adoption, enabling skills development. However, deep poverty continues to impede access to capital-intensive technologies in Africa. Moreover, infrastructure gaps like electricity shortages or digital connectivity challenges disrupt technology utilisation, though innovations are adapted for constraints (Baumüller, 2015). For example, solar-powered lanterns and basic mobile phones enable base-of-pyramid access amid underdevelopment.

Local innovation capabilities drive appropriate technology development but remain limited in Africa (Chataway et al., 2014). Foreign designs often overlook contextual needs. Grassroots maker communities prototyping locally relevant solutions are expanding through new fabrication tools. Furthermore, homegrown digital platforms also tailor services to African users. Still, most advanced technologies are imported.

Additionally, cultural and social attitudes further shape technology integration and consequences. Patriarchal norms hindering women's financial and physical mobility constrain opportunities. However, collaborative traditions facilitate the adoption of shared solutions like equipment cooperatives. Status perceptions around technologies persist but evolve with diffusion. Policy and regulatory environments must also keep pace.

Outdated bureaucracies slow tech integration while transparent, pro-innovation systems accelerate adoption. Effective policies balance innovation support, risk mitigation and social protections (Lundvall et al., 2009).

Reliable supporting infrastructure expands technology access and adoption. Electricity, roads, digital networks, supplier ecosystems, and regulatory systems enable technology acquisition and productive use. However, Africa's infrastructure lags behind other regions, straining technology integration. Insufficient electrical grid access hampers the adoption of powered tools and devices. Off-grid solar solutions bypass this barrier for applications like lighting and appliances (Amankwah-Amoah, 2015). Mini-grids also provide localised renewable energy sources. Still, unreliable electricity disrupts digital usage.

Limited Internet connectivity and bandwidth constrain engagement with online systems and digital platforms, particularly in rural areas. In contrast, urban zones enjoy greater broadband access, thus sustaining usage divides. Weak digital infrastructure risks locking African nations into outdated technical systems. However, grassroots innovations like solar microgrids, offline digital networks, and mobile systems bypass inadequate centralised infrastructure through localised solutions—thereby increasing equitable access (Olopade, 2015). For instance, small wind turbines and village hydro and biogas projects generate local renewable power. Offline data networks enable information access without full connectivity. Mobile platforms deliver services through phones. Nevertheless, significant infrastructure gaps remain. Furthermore, underdeveloped road transport infrastructure strains physical technology circulation and maintenance across Africa's vast territories. As a result, remote areas stay disconnected.

Some startups in Africa are using drones to deliver goods and services to areas with poor road infrastructure. However, many complex technologies require reliable after-sales support, such as maintenance, spare parts, and user assistance, which is often lacking in Africa. When technologies are imported without establishing local support networks, they may end up being underused or discarded. Additionally, old-fashioned and complicated government rules can make it difficult for technology entrepreneurs to introduce new products and services, slowing down adoption. However, when countries update their regulations to be clearer, more supportive of innovation, and easier to navigate, they can speed up the integration of new technologies (Aubert, 2005). Making data freely available to the public can also encourage the development of new digital solutions. Nevertheless, as policies evolve, governments must find a balance between supporting innovation and managing potential risks associated with new technologies.

High costs of acquiring and operating technologies like computers, machinery, and digital systems deter adoption in resource-constrained environments. Upfront capital investments require financing to distribute costs over time. Additionally, uncertainty over returns dampens investment in unfamiliar technologies. However, flexible purchasing models using pay-as-you-go plans and micro-loans have enabled adoption by easing cash flow barriers to acquiring digital devices and solar home systems (Winiecki & Kumar, 2014). Installment payment structures make significant capital investments affordable to low-income users. Moreover, sharing economies also reduce individual costs through cooperative equipment access models. Still, unaffordable costs severely constrain technology integration for many.

Operating expenses like Internet and electricity subscriptions also affect sustained technology use. Recurrent costs for services, maintenance, and replacements limit engagement as financial resources fluctuate. Shared infrastructure and alternating usage patterns provide cost efficiencies. Strategic adoption timing as costs decline over technology life cycles helps address affordability barriers. Cost transparency and subsidies further facilitate adoption. However, financing carries risks if repayment obligations exceed income flows, thus leading to coercive debt collection and financial distress. Therefore, responsible credit appraisal and consumer protection are essential to avoid predatory lending (Bateman, 2010). Furthermore, technology costs are highly dependent on external factors like import duties, taxes and currency fluctuations, especially given Africa's reliance on imported innovations. Local production can hedge such risks, but they remain limited. Open standards and interoperable technologies improve cost efficiency. Labour costs are a vital consideration for automation technologies. Where wages are low, automation is less compelling financially but reduces social risks. As economies develop, automation adoption may accelerate.

Local promotion by trusted community opinion leaders provides validation, while peers model positive adoption experiences. Demonstration pilots enable hands-on trials showcasing concrete utility. However, mismatches with local values and lifestyles impede the adoption of some technologies. Gender norms affect perceptions of technology utility and appropriateness. Where women have limited financial autonomy and mobility, technologies promising individual convenience are less appealing than communal solutions. Tropes of technology as a male domain further discourage female engagement.

Generational attitudes also generate adoption gaps. Youth more readily embrace fast-changing digital systems, which older citizens may see as risky or needlessly complex. However, shared community participation can bridge age-based divides. Evangelising innovation leaders should avoid alienating supporters of tradition. Perceptions evolve with exposure as risks and benefits become apparent. Initial scepticism gives way to enthusiastic adoption of beneficial technologies. However, invasive systems face growing public resistance, eroding adoption through changing attitudes.

Technologies designed for different cultural contexts frequently fail when transferred to new environments without adaptation. Relevance to localised needs, values, and operating conditions heavily influences adoption. User input guides customising systems to resonate with African realities rather than directly importing foreign models. Agricultural technologies often fail by disregarding smallholder knowledge and lacking synergy with local growing conditions (Glover et al., 2019). Imposing external solutions ignores community priorities and circumstances. Digital systems reflecting urban design preferences also underserve rural communities with modest literacy, connectivity, and power access. Consequently, weak localisation limits the success of transferred technologies. Conversely, mobile money and off-grid solar innovations tailored for affordability and usability in African contexts gained widespread adoption by addressing accessibility gaps with appropriate solutions (Onsongo & Schot, 2017). Participatory design incorporating user perspectives enables the creation of technologies that resonate with local needs.

Language and literacy constraints pose adoption barriers where technologies rely on textual interfaces. Software localised through translation expands access alongside icon-driven designs. Voice user interfaces are also more inclusive. But digital systems favouring a small minority language perpetuate exclusion. Prioritising multilingual capability and oral functions suits Africa's diverse environments. Resource limitations shape technology use patterns in the region. Despite connectivity gaps, offline digital tools meet needs. Shared access models overcome high individual costs. Pay-as-you-go financing spreads capital investments over time (Rolffs et al., 2015). Optimisation for unreliable infrastructure broadens reach. But deep poverty still restricts adoption. Additionally, cultural values further inform adaptation needs. Privacy controls address data concerns.

Usage restrictions respond to harmful content fears. Gender sensitivity counters patriarchal constraints. While socially aligned innovations speed adoption, some changes remain contested.

Technology adoption varies significantly across African demographic segments due to uneven access to resources, infrastructure, digital literacy, and value perceptions. Cost barriers severely constrain technology use by low-income populations. Rural dwellers lag behind urban zones in acquiring grid-dependent devices. Gender divides stem from women's more limited education, financial autonomy and mobility. In contrast, youth more readily embrace new technologies through peer exposure. Targeted policies help democratise adoption by addressing community-specific barriers (see Asongu & Boateng, 2018). However, poverty continues to block digital access and technology ownership for a large section of Africa's population. Consumers face structural disadvantages at the bottom of the pyramid in affording and optimally utilising technologies designed for more privileged users. Incremental financing models like pay-as-you-go solar power widen adoption by easing cash flow constraints.

Significant rural-urban divides persist in technology adoption across Africa, as modern technologies spread faster in urban areas than in rural regions. This urban concentration reflects better infrastructure, electricity access, income levels, and human capital in cities, enabling technology acquisition and use. Higher educated and literate populations in urban areas also facilitate optimal utilisation of technologies. Rural-urban gaps arise from poorer electricity and Internet connectivity infrastructure in remote areas, thus constraining powered device usage and digital platform engagement (Hasbi & Dubus, 2020). Weaker educational resources also contribute to lower digital literacy in rural zones. However, grassroots innovations like village solar microgrids, offline digital networks, and mobile phone systems help decentralise access and bypass centralised infrastructure limitations through localised solutions (Ockwell & Byrne, 2017). Still, most technology development and targeting focuses on urban contexts while underserving rural needs.

3.3 TECHNOLOGY TRANSFER AND DIFFUSION PATTERNS

Technology adoption and spread across Africa has been uneven, with significant differences across countries, urban-rural divides, gender and income groups, and regions. Studies have found faster technology uptake in middle-income African countries than in low-income nations (Albiman

& Sulong, 2016). Moreover, pronounced urban-rural digital divides exist within countries, with modern tech concentration in cities (Haftu, 2019). Gender gaps arise from women's lower digital access owing to financial and mobility limitations in patriarchal cultures (Asongu & Odhiambo, 2018). Income levels strongly correlate with technology ownership, as costs bar the bottom of the pyramid of users (Shambare, 2014). Furthermore, regional variations are visible based on colonial legacies, trade openness, infrastructure availability, and competitiveness.

Uneven spread and adoption gaps are evident between countries and within countries. Studies find faster adoption of technologies like mobile phones and the Internet in middle-income countries compared to low-income countries in Africa (Albiman & Sulong, 2016; Baliamoune-Lutz, 2003). Significant digital divides persist between African countries depending on economic development levels (IMF, 2020). For instance, Internet penetration in 2019 was very high in Seychelles and Morocco but very low in Somalia and Eritrea (IMF, 2020). Urban-rural digital divides are substantial, as modern technologies spread faster in cities than in rural areas. Urban residents have over five times more mobile phone access than rural populations in some African countries (Haftu, 2019). Internet adoption in Africa is primarily concentrated in a few urban zones like Lagos, Cairo, Cape Town, and Johannesburg, while it remains low in rural areas. Infrastructural limitations and lack of electricity hamper technology access in rural Africa.

Pronounced gender divides exist in technology adoption across Africa, with men exhibiting higher access and usage rates than women. For instance, in poorer countries like Uganda, women are 50% less likely to own mobile phones than men (Asongu & Odhiambo, 2018). These adoption gaps arise from women's more constrained financial autonomy, digital skills, literacy, and physical mobility due to prevailing patriarchal gender norms and limitations in male-dominated African societies. Socio-cultural restrictions also hamper women's access to technology. Additionally, as discussed above, tropes of technology as a male domain discourage female technology engagement.

Technology adoption patterns in Africa correlate strongly with income levels, as significant divides persist between richer and poorer population segments. Mobile phone and Internet access remain far lower among low-income groups than wealthier consumers (Asongu & Odhiambo, 2018). The high costs of devices, network subscriptions, services, and data constitute significant barriers to technology acquisition and use by bottom-of-the-pyramid users who struggle to afford basic needs. Financial constraints also explain limited technology integration among micro and informal enterprises that dominate Africa's private sector (Shambare, 2014). These income divides reflect structural economic disadvantages and exclusion faced by low-income people, along with innovation ecosystems and policies favouring elite privileges over marginalised populations.

Regional variations are visible in technology diffusion across Africa based on colonial legacies, geographical proximity to innovating regions, trade openness, competitiveness, infrastructure availability, and regulatory systems. With better connectivity, northern and Southern African countries like Egypt, Kenya, and South Africa have faster ICT uptake than West Africa or landlocked nations like Chad and Niger with lower global integration (Albiman & Sulong, 2016). Small island economies like Mauritius and Cape Verde had some of the highest mobile and Internet penetration in the mid-2000s, facilitated by solid telecom liberalisation and competition policies (Baliamoune-Lutz, 2003).

Urban areas in Africa witness much faster technology uptake than rural regions due to better infrastructure, electricity access, income levels, and human capital in cities facilitating adoption. Higher literate and educated populations in cities also enable a greater ability to acquire and optimally utilise modern technologies. For instance, Internet penetration in Africa was over three times higher in top urban centres than in rural zones during 1998–2005 (Myovella et al., 2020). The urban-rural divide was even starker in Ghana, with 15 times higher Internet adoption in cities compared to rural areas as of 2013. Bridging these stark adoption gaps requires improving rural electricity, connectivity, education, and affordability. But grassroots informal innovations like village solar microgrids and offline digital networks can also decentralise access more equitably. Telecenters supplying public infrastructure in rural areas further enable technology exposure.

Africa's significant income inequalities sustain pronounced technology adoption gaps between richer and poorer populations. At the bottom of the pyramid, low-income communities still have negligible access to digital technologies compared to the middle and upper-income segments. High costs of devices, network subscriptions, and data put modern ICTs out of reach for most low-income households who struggle to afford basic needs (Gillwald et al., 2018). Small informal microenterprises face greater financial, skill, and infrastructure barriers in adopting digital platforms than larger formal firms. These income divides result from structural economic disadvantages faced by people experiencing poverty, innovation ecosystems, and policy frameworks favouring elite interests over marginalised populations.

Regional diffusion patterns for technologies like mobile and Internet vary across Africa based on colonial influences, proximity to innovating centres, integration in global trade and production networks, geographical characteristics, regulatory approaches, and competitiveness. North African countries like Egypt, with historic ties to Europe and West Asia, have advanced further in ICT networks. Southern Africa is also more technologically advanced, benefiting from links to South Africa, leading to digital integration (Myovella et al., 2020). Landlocked and isolated countries in central Africa with lower international engagement have lagged in adoption. Island states like Mauritius, with vital services sectors, adopted earlier due to conducive regulations attracting telecom investment and competition. Resource-rich countries also focused more on extractive industries rather than developing digital capabilities.

One crucial factor shaping technology diffusion patterns in Africa is the role of international trade and foreign direct investment (FDI). Countries with higher levels of trade openness and FDI inflows tend to experience faster technology adoption, as these channels facilitate the transfer of knowledge, skills, and advanced equipment (Osabutey & Okoro, 2015). Multinational corporations investing in African markets often bring cutting-edge technologies and production processes, which can spill over to domestic firms through linkages, imitation, and labour mobility (Boly et al., 2014). However, the extent of technology transfer through FDI varies across sectors. It depends on the absorptive capacity of local firms, highlighting the importance of policies to strengthen human capital and innovation capabilities (Osabutey & Okoro, 2015).

Another emerging trend is the growing adoption of frugal innovations and bottom-of-the-pyramid (BOP) solutions tailored to the needs and constraints of low-income African consumers. Frugal innovations are cost-effective, easy-to-use, and resilient products or services that meet the essential needs of resource-constrained populations (Radjou & Prabhu, 2015). Examples include mobile money services, solar lanterns, and water purification systems. By leveraging local knowledge, resources, and distribution networks, frugal innovations can reach underserved markets and contribute to inclusive development (Knorringa et al., 2016). However, scaling up frugal innovations often requires supportive ecosystems, including access to finance, market information, and regulatory frameworks that incentivise pro-poor solutions (Radjou & Prabhu, 2015).

The rise of digital platforms and the sharing economy also transforms technology adoption patterns in Africa. Digital platforms like mobile apps, e-commerce marketplaces, and social media networks enable African consumers and businesses to access a broader range of goods, services, and information (David-West et al., 2018). Sharing economy models, such as ride-hailing apps and co-working spaces, optimise idle assets and create new livelihood opportunities in urban centres (Yeboah-Asiamah et al., 2016). However, the growth of digital platforms also raises concerns about data privacy, algorithmic bias, and the precarity of gig work, calling for regulatory frameworks that balance innovation with consumer protection and labour rights (David-West et al., 2018).

3.4   Impacts and Unintended Consequences

Technology adoption in Africa has yielded a mix of development benefits and adverse socioeconomic effects, externalities and ethical risks requiring balanced policy approaches. On the positive side, studies show linkages between technology adoption and higher economic growth, increased financial inclusion, expanded educational access and health service improvements in parts of Africa (Albiman & Sulong, 2016; Asongu & Odhiambo, 2018). However, concerns exist around growing inequality from digital divides and automation job losses, cultural change risks from unfiltered social media proliferation, fraud/crime issues with digital finance, environmental pollution from e-waste dumping, public health threats like Internet addiction, and data ethics challenges around digital surveillance and AI biases (Lekhanya, 2013; Park & Choi, 2019).

Positive links between technology and development in Africa, including economic growth, financial inclusion, educational access and health improvements, are beyond question. Mobile phone adoption contributed over 0.5 percentage points to per capita GDP growth in Africa from 1998 to 2014, facilitating communication, market access, and efficiency for firms (Albiman & Sulong, 2016). Moreover, digital financial services enable financial inclusion for the unbanked population through mobile money (Asongu & Odhiambo, 2018). Educational technologies are expanding access and improving pedagogy, as seen in university adaptations of e-learning, though uptake varies across countries. Telemedicine and mHealth initiatives enhance healthcare access in remote areas. Digital platforms support entrepreneurship and the informal sector (Lekhanya, 2013). However, concerns exist around negative socioeconomic impacts. Automation threatens to displace workers and widen inequalities as low-skilled jobs are most vulnerable (Amankwah-Amoah et al., 2018). The data divide excludes marginalised groups from digital opportunities (Asongu & Odhiambo, 2018). Increased use of social media and digital entertainment raises fears regarding cultural influence and loss of local identities (Lekhanya, 2013), besides concerns over digital addiction and isolation. Financial fraud and cybercrime risks are proliferating with digitalisation, though awareness and safeguards remain low. High imports of digital technologies constrain local manufacturing and perpetuate the technological dependence of African economies on advanced nations.

Growing electronic waste accumulation in Africa from short life cycles of devices like mobile phones and computers leads to significant soil, air and water pollution through hazardous chemicals in informal dumping grounds with limited safe disposal (Park & Choi, 2019). Furthermore, increased use of digital devices raises exposure to radiation, eyestrain and other musculoskeletal disorders as awareness of healthy technology usage habits remains low in the region. Internet and social media addiction enhanced by expanding digital access leads to unintended public health risks, including physical inactivity, depression, anxiety, and sleep disorders among children and youth. However, environmental and health impact assessments regarding the potentially detrimental consequences of accelerating technology adoption are still lacking in Africa. Promoting appropriate e-waste recycling, radiation safety, screen time limits, and digital well-being through regulations and awareness campaigns can help mitigate emerging issues. A precautionary approach is prudent for managing risks.

Ethical considerations around data privacy, AI biases, labour displacement, and addiction call for caution in African technology adoption. Weak data protections raise privacy abuse issues and may be seen in an unauthorised sharing of mobile subscribers' personal information. Risks of gender, racial, and income biases exist in AI systems developed using limited local context and non-representative data. Automation and digitisation need responsible transitions for workforce adjustments. And digital addictions among youth merit preventive interventions. Evidence on technology's developmental impact shows benefits like economic growth, inclusion, and improved services, weighed against inequality risks, dependence, socio-cultural change, and health issues. A nuanced regulatory approach balancing innovation support, risk mitigation, local context adaptation, and responsible use can optimise positive outcomes from Africa's adoption of technology.

Digital technologies contributed to higher GDP growth, employment creation in services sectors, increased productivity and exports, and financial inclusion for marginalised groups in Africa (see Albiman & Sulong, 2016; Asongu & Odhiambo, 2018). Mobile money enabled access to transfers, savings, credit, and other financial services for the unbanked population. Educational technologies are improving teaching methods and learning outcomes, enabling remote learning during COVID-19 closures. Telemedicine helped expand healthcare access to underserved communities through mobile apps allowing remote consultations and diagnosis. Digital platforms supported informal entrepreneurship and SME growth in African countries (Lekhanya, 2013). However, increased automation and digitisation raise concerns over job losses and growing inequality, as low-skilled workers are most vulnerable to displacement (Amankwah-Amoah et al., 2018). The data divide excludes youth, women, and rural and poorer populations from fully benefiting from digital opportunities, thus reproducing broader inequalities. Cultural critics argue that social media and expanding digital content lead to foreign cultural domination and loss of local identities, besides risks of digital addiction and social isolation. Given low cybersecurity awareness, digital financial services enable cyber fraud and scams, leading to user losses. High imports of digital technologies constrain domestic manufacturing sector development in Africa, entrenching dependence on foreign innovations.

Unintended health consequences include electronic waste accumulation and environmental pollution from hazardous device chemicals (Asongu & Nwachukwu, 2017). The growing use of computers, mobile phones, and digital platforms also increases exposure to radiation, eyestrain, and other disorders as usage guidelines remain limited. Internet addiction enhances risks of physical inactivity, depression, and anxiety among children and youth. Ethical considerations around Africa's technology adoption include data privacy concerns given weak protections, biases in automated decision-making systems, especially gender and racial preferences, workforce displacement from automation requiring responsible transition policies, and digital addiction issues needing preventive interventions (Wakunuma & Masika, 2017).

The growth of e-commerce and online platforms in Africa is creating new opportunities for entrepreneurs and small businesses but also raising concerns about market concentration and the dominance of foreign tech giants. While platforms like Jumia and Amazon are expanding access to regional and global markets for African sellers, they also risk crowding out local competitors and perpetuating dependency on external technology providers (Bukht & Heeks, 2017).

The increasing adoption of artificial intelligence (AI) and automation technologies in Africa poses opportunities and risks for labour markets and social equity. On one hand, AI-powered innovations in sectors like agriculture, healthcare, and education could help tackle development challenges and improve service delivery (Smith & Neupane, 2018). On the other hand, the automation of low-skilled jobs in manufacturing and services could lead to technological unemployment and worsen inequality, particularly for women and youth (Ndung'u & Signé, 2020).

The rapid spread of social media and messaging apps in Africa transforms social interactions, political participation, and information flows. Platforms like Facebook, WhatsApp, and Twitter have become powerful tools for mobilising citizens, holding governments accountable, and amplifying marginalised voices. However, they have also been misused to spread misinformation, hate speech, and political propaganda, potentially destabilising democratic processes and social cohesion (Ndlela & Mano, 2020).

3.5   STRATEGIES FOR INCLUSIVE INNOVATION ECOSYSTEMS

Fostering inclusive technology innovation and adoption in Africa involves embracing user-centred design, community self-management models, grassroots maker communities, public digital access, and social learning platforms. User-centred and participatory design ensures that solutions resonate with local priorities and contexts. Community oversight for technology access and maintenance through cooperative structures boosts sustainability. Moreover, promoting grassroots entrepreneurship and fabrication spaces enables affordable localisation. Public digital centres provide shared infrastructure for peer learning and skills acquisition. Furthermore, social learning leverages networks for the informal diffusion of knowledge and know-how. Such strategies can make innovation more bottom-up, equitable and empowering. Mainstreaming inclusive design thinking is vital, given Africa's diversity. No single solution will suit all communities. Regular user engagement and feedback help match offerings to evolving needs. Communities should lead technology integration on their own terms. Externally imposed systems often fail without local ownership. Broadening opportunities for ordinary Africans to shape solutions tailored to their realities is imperative for responsible innovation (Kraemer-Mbula & Wamae, 2010; Lundvall et al., 2009).

As I will argue in Chap. 6, user-centred design ensures technologies are customised for local priorities and contexts through co-design with target users. For instance, involving farmer groups in designing solutions can improve localisation for smallholder needs and environments (Meijer et al., 2015). Participatory approaches like crowdsourcing user feedback help make innovations more accessible and intuitive for diverse populations, including those with limited education or digital skills. Community oversight and self-management of technologies through cooperative structures facilitate adoption and sustainability. Farmer cooperatives managing shared farm machinery ensure affordable access, maintenance, and maximum utilisation through scheduling systems. Community technology centres providing Internet access run as local associations or partnerships, combining decentralisation with collaborative problem-solving (Wamuyu, 2015).

Promoting grassroots entrepreneurship and maker spaces enables locality-relevant solutions. Technology hubs and maker spaces with shared fabrication tools support the prototyping of localised innovations addressing accessibility, cost, and infrastructure issues (Amankwah-Amoah, 2015). Government and donor agencies are backing programmes to help youth develop affordable digital innovations tailored to their communities. Public access facilities like community technology centres provide shared digital resources, allowing citizens to learn by doing. Rural knowledge centres with Internet connectivity and mentors help farmers, microentrepreneurs, and students acquire digital skills through hands-on practice (Gyamfi et al., 2017). Public libraries also supplement school-based ICT education through digital literacy programmes. Peer learning approaches leverage social networks for informal knowledge diffusion. Farmer field schools enable peer-to-peer transfer of technology know-how through observation, experimentation, and exchange of experiences. Such complementary channels are critical where formal technical skills and support are limited.

Strengthening intellectual property rights (IPR) regimes and technology transfer policies can help foster indigenous innovation and encourage foreign investment in R&D in Africa. Well-designed IPR systems that balance the interests of inventors and the public's interests can incentivise local firms to invest in developing and commercialising new technologies (Asongu & Boateng, 2018). At the same time, technology transfer requirements for foreign investors, such as local content policies and joint venture partnerships, can facilitate knowledge spillovers and build local innovation capabilities (Osabutey & Croucher, 2018). However, such policies must be carefully crafted to avoid deterring FDI or limiting access to essential technologies.

Promoting open innovation and collaboration between academia, industry, and government can help accelerate the development and diffusion of locally relevant solutions. Open innovation models that leverage external knowledge sources and engage diverse stakeholders can speed up the innovation process and improve the contextual fit of new technologies (Egbetokun et al., 2017). University-industry linkages, such as technology transfer offices, incubators, and research parks, can facilitate the commercialisation of academic research and provide students with practical innovation skills (Zavale & Langa, 2018). Triple-helix partnerships between government agencies, private firms, and research institutions can align innovation agendas with national development priorities and pool resources for high-impact projects (Etzkowitz & Zhou, 2018).

Investing in digital infrastructure and skills development can help bridge the digital divide and enable more Africans to participate in the innovation economy. Expanding access to affordable broadband Internet, reliable electricity, and digital devices is essential for unlocking the potential of digital technologies for education, entrepreneurship, and service delivery (Hasbi & Dubus, 2020). Integrating ICT skills into school curricula, vocational training programmes, and lifelong learning opportunities can help build a digitally literate workforce and user base (Howard, 2023). Particular attention should be given to closing gender gaps in digital access and skills, as well as supporting marginalised groups such as persons with disabilities and rural communities. Investing in local language content, user-friendly interfaces, and assistive technologies can help make digital innovations more inclusive and accessible.

3.6   Conclusion

Technology adoption and integration pathways in Africa have been shaped by complex, multidimensional factors encompassing the availability of localised resources and capabilities, attitudes and priorities of intended user communities, relevance to cultural contexts, demographic differences, and considerations around infrastructure and costs. Indeed, the uneven diffusion outcomes and digital divides reflect these community-specific determinants. Therefore, achieving more sustainable, responsible, and inclusive technology futures in Africa requires going beyond techno-cratic models to integrative, participatory approaches. These should be grounded in understanding diverse social, economic, and cultural enablers and barriers across the continent's varied locales. Imposing external solutions without adaptation risks failure and disempowerment.

The chapter has highlighted the need for fundamental reforms in innovation ecosystems to centre ethical assessment, user empowerment, and appropriate design. Participatory processes engaging ordinary citizens, civil society groups and policymakers are vital to steer technology trajectories in socially progressive directions. Moreover, investments in human capabilities and inclusive digital infrastructure should accompany technology initiatives to democratise opportunities. Continued interdisciplinary research on balancing emerging technologies' benefits and risks can guide evidence-based governance. However, communities must lead integration on their terms to meet localised priorities. Technology designed for Africa, not just deployed in Africa, is imperative for this transformative yet responsible development.

Several additional factors and trends are shaping technology diffusion patterns in Africa, such as the role of international trade and FDI in facilitating knowledge transfer, the rise of frugal innovations and digital platforms tailored to low-income consumers, and the growing adoption of the sharing economy in urban centres. These developments present opportunities and challenges for inclusive innovation, requiring policies that balance technology access with data privacy, consumer protection, and labour rights. The impacts and unintended consequences of emerging technologies in Africa also demand attention. The growth of e-commerce platforms is creating new market opportunities but also raising concerns about foreign dominance and crowding out local competitors. The increasing adoption of AI and automation poses risks of technological unemployment and worsening inequality, particularly for women and low-skilled workers. The rapid spread of social media is transforming citizen participation and information flows and enabling the spread of misinformation and hate speech. Navigating these complex trade-offs will require proactive governance frameworks and multi-stakeholder collaboration.

To foster inclusive innovation ecosystems, African countries should prioritise strategies such as strengthening IPR regimes and technology transfer policies, promoting open innovation and academia-industry-government collaboration, investing in digital infrastructure and skills development, and supporting local language content and user-friendly interfaces. Particular attention should be given to closing gender gaps in digital access and skills, as well as enabling the participation of marginalised groups such as persons with disabilities and rural communities. With inclusive innovation and enabling ecosystem reforms, technology adoption patterns can shift from the imposition of inconsistent systems towards integrating appropriate, empowering solutions designed for equitable advancement. The full promise of technologies resonating with Africa's diverse realities can be realised through cooperative action by governments, businesses, communities, and researchers.

REFERENCES

Aker, J. C., & Mbiti, I. M. (2010). Mobile phones and economic development in Africa. *Journal of Economic Perspectives, 24*(3), 207–232.

Albiman, M. M., & Sulong, Z. (2016). The role of ICT in the economic growth in the Sub-Saharan African region (SSA). *Journal of Science and Technology Policy Management, 7*, 306–329.

Amankwah-Amoah, J. (2015). Solar energy in sub-Saharan Africa: The challenges and opportunities of technological leapfrogging. *Thunderbird International Business Review, 57*(1), 15–31.

Amankwah-Amoah, J. (2016). The evolution of science, technology and innovation policies: A review of the Ghanaian experience. *Technological Forecasting and Social Change, 110*, 134–142.

Amankwah-Amoah, J., Osabutey, E. L., & Egbetokun, A. (2018). Contemporary challenges and opportunities of doing business in Africa: The emerging roles and effects of technologies. *Technological Forecasting and Social Change, 131*, 171–174.

Asongu, S., & Nwachukwu, J. (2017). *Mobile phone innovation and environmental sustainability in Sub-Saharan Africa.* African Governance and Development Institute WP/17/025. https://doi.org/10.2139/ssrn.2999823

Asongu, S. A., & Boateng, A. (2018). Introduction to the special issue: Mobile technologies and inclusive development in Africa. *Journal of African Business, 19*(3), 297–301.

Asongu, S. A., & Odhiambo, N. M. (2018). ICT, financial access and gender inclusion in the formal economic sector: Evidence from Africa. *African Finance Journal, 20*(2), 45–65.

Asongu, S. A., & Odhiambo, N. M. (2019). Challenges of doing business in Africa: A systematic review. *Journal of African Business, 20*(2), 259–268.

Aubert, J. E. (2005). *Promoting innovation in developing countries: A conceptual framework* (World Bank Policy Research Working Paper, 3554).

Baliamoune-Lutz, M. (2003). An analysis of the determinants and effects of ICT diffusion in developing countries. *Information Technology for Development, 10*(3), 151–169.

Bateman, M. (2010). *Why doesn't microfinance work?: The destructive rise of local neoliberalism*. Zed Books Ltd.

Baumüller, H. (2015). Assessing the role of mobile phones in offering price information and market linkages: The case of M-Farm in Kenya. *The Electronic Journal of Information Systems in Developing Countries, 68*(1), 1–16.

Boly, A., Coniglio, N. D., Prota, F., & Seric, A. (2014). Diaspora investments and firm export performance in selected sub-Saharan African countries. *World Development, 59*, 422–433.

Bukht, R., & Heeks, R. (2017). *Defining, conceptualising and measuring the digital economy*. (Development Informatics Working Paper No. 68). University of Manchester.

Chataway, J., Hanlin, R., & Kaplinsky, R. (2014). Inclusive innovation: An architecture for policy development. *Innovation and Development, 4*(1), 33–54.

David-West, O., Iheanachor, N., & Umukoro, I. (2018). Sustainable business models for the creation of mobile financial services in Nigeria. *Journal of Innovation and Entrepreneurship, 7*(1), 1–16.

Egbetokun, A., Oluwadare, A. J., Ajao, B. F., & Jegede, O. O. (2017). Innovation systems research: An agenda for developing countries. *Journal of Open Innovation: Technology, Market, and Complexity, 3*(1), 25.

Etzkowitz, H., & Zhou, C. (2018). Innovation incommensurability and the science park. *R&D Management, 48*(1), 73–87.

Gillwald, A., Odufuwa, F., & Mothobi, O. (2018). *The state of ICT in Nigeria*. Research ICT Africa.

Glover, D., Sumberg, J., Ton, G., Andersson, J., & Badstue, L. (2019). Rethinking technological change in smallholder agriculture. *Outlook on Agriculture, 48*(3), 169–180.

Gyamfi, G. D., Modjinou, M., & Djordjevic, S. (2017). Improving electricity supply security in Ghana—The potential of renewable energy. *Renewable and Sustainable Energy Reviews, 43*, 1035–1045.

Haftu, G. G. (2019). Information communications technology and economic growth in Sub-Saharan Africa: A panel data approach. *Telecommunications Policy, 43*(1), 88–99.

Hasbi, M., & Dubus, A. (2020). Determinants of mobile broadband use in developing economies: Evidence from Sub-Saharan Africa. *Telecommunications Policy, 44*(5), 101944.

Howard, C. (2023). *Digital skills for youth employment in Africa*. Include Knowledge Platform on Inclusive Development Policies.

IMF. (2020). *Regional economic Outlook Sub-Saharan Africa: COVID-19: An unprecedented threat to development.* International Monetary Fund.

Knorringa, P., Peša, I., Leliveld, A., & Van Beers, C. (2016). Frugal innovation and development: Aides or adversaries? *The European Journal of Development Research, 28*(2), 143–153.

Kraemer-Mbula, E., & Wamae, W. (Eds.). (2010). *Innovation and the development agenda.* OECD Publishing.

Lekhanya, L. M. (2013). The use of social media and social networks as the promotional tool for rural small, medium and micro enterprises in KwaZulu-Natal. *International Journal of Scientific and Research Publications, 3*(7), 1–7.

Lundvall, B. Å., Joseph, K. J., Chaminade, C., & Vang, J. (Eds.). (2009). *Handbook of innovation systems and developing countries: Building domestic capabilities in a global setting.* Edward Elgar Publishing.

Meijer, S. S., Catacutan, D., Ajayi, O. C., Sileshi, G. W., & Nieuwenhuis, M. (2015). The role of knowledge, attitudes and perceptions in the uptake of agricultural and agroforestry innovations among smallholder farmers in sub-Saharan Africa. *International Journal of Agricultural Sustainability, 13*(1), 40–54.

Myovella, G., Karacuka, M., & Haucap, J. (2020). Digitalisation and economic growth: A comparative analysis of Sub-Saharan Africa and OECD economies. *Telecommunications Policy, 44*(2), 101856.

Ndlela, M. N., & Mano, W. (Eds.). (2020). *Social media and elections in Africa, Volume 1: Theoretical perspectives and election campaigns.* Springer.

Ndung'u, N., & Signé, L. (2020). *The fourth industrial revolution and digitisation will transform Africa into a global powerhouse. Foresight Africa 2020.* Brookings Institution.

Ockwell, D., & Byrne, R. (2017). *Sustainable energy for all: Innovation, technology and pro-poor green transformations.* Routledge.

Olopade, D. (2015). *The bright continent: Breaking rules and making change in modern Africa.* Houghton Mifflin Harcourt.

Onsongo, E. K., & Schot, J. (2017). *Inclusive innovation and rapid sociotechnical transitions: The case of mobile money in Kenya* (SPRU Working Paper Series (SWPS), 2017-07).

Osabutey, E. L., & Croucher, R. (2018). Intermediate institutions and technology transfer in developing countries: The case of the construction industry in Ghana. *Technological Forecasting and Social Change,* pp. 128, 154–163.

Osabutey, E. L., & Okoro, C. (2015). Political risk and foreign direct investment in Africa: The case of the Nigerian telecommunications industry. *Thunderbird International Business Review, 57*(6), 417–429.

Park, H., & Choi, S. O. (2019). Digital innovation adoption and its economic impact focused on path analysis at national level. *Journal of Open Innovation: Technology, Market, and Complexity, 5*(3), 56. https://doi.org/10.3390/joitmc5030056

Radjou, N., & Prabhu, J. (2015). *Frugal innovation: How to do more with less.* British Books for Managers (BBM).

Rolffs, P., Ockwell, D., & Byrne, R. (2015). Beyond technology and finance: Pay-as-you-go sustainable energy access and theories of social change. *Environment and Planning A: Economy and Space, 47*(12), 2609–2627.

Shambare, R. (2014). The adoption of WhatsApp: Breaking the vicious cycle of technological poverty in South Africa. *Journal of Economics and Behavioral Studies, 6*(7), 542–550.

Smith, M. L., & Neupane, S. (2018). *Artificial intelligence and human development: Toward a research agenda.* International Development Research Centre.

Wakunuma, K. J., & Masika, R. (2017). Cloud computing, capabilities and intercultural ethics: Implications for Africa. *Telecommunications Policy, 41*(7-8), 695–707.

Wamuyu, P. K. (2015). The impact of information and communication technology adoption and diffusion on technology entrepreneurship in developing countries: The case of Kenya. *Information Technology for Development, 21*(2), 253–280.

Winiecki, J., & Kumar, K. (2014). *Access to energy via digital finance: Overview of models and prospects for innovation.* CGAP.

Yeboah-Asiamah, E., Quaye, D. M., & Nimako, S. G. (2016). The effects of lucky draw sales promotion on brand loyalty in the mobile telecommunication industry. *African Journal of Economic and Management Studies, 7*(1), 109–123.

Zavale, N. C., & Langa, P. V. (2018). University-industry linkages' literature on Sub-Saharan Africa: Systematic literature review and bibliometric account. *Scientometrics, 116*(1), 1–49.

CHAPTER 4

Technologising Tradition or Traditionalising Technology?

4.1   INTRODUCTION

Africa's diverse artistic, linguistic, social, and knowledge traditions face complex transitions amid accelerating technological change. As digital tools and paradigms spread across the continent, thoughtful negotiation is required to integrate them with enduring cultural fabrics. Foreign technologies designed without considering local contexts risk undermining rather than uplifting Africa's rich heritage. Dominant narratives of 'modernisation' and 'development' imposed from outside have long portrayed African knowledge systems as 'primitive' and 'backwards' to justify their replacement under colonial civilising missions and post-independence agendas (Baker, 2012). However, in light of today's multifaceted changes, appreciating African philosophies and cosmologies is crucial for guiding appropriate adaptations that sustainably blend old and new.

This chapter examines the contingent, multilayered ways external technologies intersect with Africa's intangible and embodied cultural dimensions. Specifically, it explores how African orature, lingua culture, aesthetics, social institutions, and communal ethics face both creative opportunities and problematic disruptions amid rapid technologisation. On the one hand, ruptures of heritage occur through decontextualised appropriation or paradigm displacement; on the other hand, synergies emerge from thoughtful integration (Kaschula & Mostert, 2009). Nevertheless, Africa's immense diversity cautions against

Y. Ndasauka, *African Mind, Culture, and Technology*,
https://doi.org/10.1007/978-3-031-62979-2_4

techno-determinist assumptions of unidirectional cultural loss. Thus, careful philosophical analysis of the varied impacts is necessary to reveal the need for strengthening localisation and self-determination.

For centuries, external paradigms have dismissed African knowledge systems, arts, and values as 'primitive' to justify their replacement through civilising missions under colonialism and modernisation agendas post-independence (Jewsiewicki, 1989). Consequently, anthropological digital archives and databases have frequently extracted cultural artefacts as scientific specimens or exotic objects detached from living communities (Brah & Coombes, 2000). Moreover, biopiracy removes indigenous knowledge from communal control (Mgbeoji, 2006). In this context, foreign technologies deceptively appear as apolitical tools promising progress by mimicking the West. However, today's complex changes reveal the risk of eroding Africa's robust knowledge foundations if modernisation discourses obscure local contexts and agency. For instance, orature faces creative possibilities and disruptions from decontextualised digitisation platforms that often ignore customary protocols (Asuquo et al., 2023). Similarly, the unreflective importing of Western education and media systems displaces situated, embodied learning vital for intergenerational knowledge transmission and the socialisation of youth on African terms (see Zeleza, 1997). As a result, externally designed technologies often fracture the holistic relationships between knowledge, practice, identity, and environments rooted in African philosophies.

Examples also reveal the thoughtful integration of tools to sustain collective advancement and cultural dynamism by augmenting traditional mediums with expanded reach. For instance, collaborative digitisation amplifies marginalised local voices over silencing them (Taylor, 2016). Likewise, platforms fostering two-way dialogue deepen engagement with living heritages, while multilingual websites and translators enable vernacular visibility and exchange online. Artists fuse ancestral motifs with emerging materials to sustain symbolic resonance. When appropriately deployed for public service, technologies expand knowledge access and networks across borders. However, risks of cultural imperialism through purportedly neutral tools remain. Mindfulness is needed to strengthen communal control over if, when, and how technologies interface with African worlds. Locating agencies within African communities to direct change processes is vital for transitioning from disruptive modernisation to innovative sustainability in Africa. Nevertheless, dominant assumptions of technology as culturally neutral persist, obscuring displacement risks. Rethinking innovation requires centring African philosophies of collective advancement, sacred balance, and human dignity, often marginalised as 'traditional' by Eurocentric paradigms (Owusu-Ansah & Mji, 2013). Valuing communal over individualistic values provides grounding to assess and shape technological integration in ways that protect shared heritages central to identity.

This chapter takes a deep dive into the complex interplay between modern technologies and the rich tapestry of African cultural heritage. The next sections explore the intersection of technology with various dimensions of African culture, including oral traditions, language and communication, artistic expression, aesthetics, and social institutions and practices. The sections present a nuanced analysis of the creative possibilities and disruptive tensions that arise as digital platforms and tools interact with these cultural domains. The chapter paints a complex picture of Africa's technocultural landscapes, from the potential of technology to support language revitalisation efforts and preserve oral traditions to the risks of decontextualisation, commodification, and erosion of embodied knowledge. Importantly, it emphasises the diversity of African cultures and the agency of local communities in navigating technological change, arguing against essentialist narratives. The conclusion ties together these themes, calling for a decolonial approach to innovation that centres on African epistemic resources, communal ethics, and holistic well-being, ultimately envisioning a future where African wisdom guides the repurposing of socio-technical systems to advance human flourishing in the digital age.

4.2   Technology's Intersection with Oral Traditions

Africa possesses a resilient heritage of oral knowledge traditions, including folklore, epic narrations, ritual poetry, mythologies, and other cultural forms conveyed intergenerationally through embodied storytelling performance and apprenticeship. However, the rapid spread of modern digital technologies and global communications media intersecting with these enduring oral arts has generated complex tensions and trade-offs between sustaining living heritage and enabling new modes of cultural expression, dissemination, and evolution. On the one hand, information and communication technologies present innovative possibilities to maintain and build upon oral traditions using tools like collaborative archives, multimedia formats, and digital platforms that can potentially democratise access, expand creativity, and enrich interactivity around traditional knowledge (Kaschula & Mostert, 2009). On the other hand, critics caution that uncritical, decontextualised applications of technology absent traditional safeguards could lead to problematic erosion or commodification of oral heritage by disrupting communal rituals, protocols, and worldviews evolved over generations to guide knowledge exchange in African communities (Akinwale, 2012; Ngulube, 2002).

Careful, thoughtful integration of appropriate digital tools that respect and amplify cultural fabrics can catalyse positive synergy between oral wisdom and emerging media. Two-way, participatory platforms enabling grassroots engagement and showcasing performer agency have fostered cultural continuation (Akinwale, 2012). However, linear assumptions that oral traditions must necessarily modernise through technology risk undervaluing the richness of living heritage and imposing disruptive foreign paradigms unsuited to diverse African realities. Moreover, African orature's communal, embodied, and spiritual dimensions do not always readily translate into digital formats designed for Western knowledge systems (Mawere & Mubaya, 2016).

Orality in Africa serves essential functions beyond information transfer, encoding wisdom, identity, ethics, and ontologies through embodied performance arts adapted locally over centuries (Finnegan, 2012). Understanding this complexity is vital. Multi-sensory orature rituals retain relevance for social cohesion and meaning-making and are not easily replicated in digital formats (Kaschula, 2004). Sophisticated protocols, rituals, and accountability systems govern intergenerational knowledge sharing and verification maintained through performance arts and mentorship (Ngulube, 2002). Native orality reflects and transmits communal values, situating knowledge relationally rather than individualistically (Mawere & Mubaya, 2016).

Rapid digitisation and globalisation thus engender complex change processes with mixed outcomes. Tools offer valuable opportunities where performers innovatively adapt technologies to expand intergenerational transmission on their terms. For example, digital storytelling platforms enabling grassroots participation and recognising performer agency can foster creative cultural continuity. However, critics note that online flows prioritising Anglophone content and paradigms risk eroding linguistic diversity and ritual foundations of orature. The imposition of external logic of monetisation, standardisation, and efficiency often undermines protocols for sharing wisdom responsibly (Akinwale, 2012). Decontextualised digitisation that fragments holistic knowledge systems and disrupts communal safeguards presents cultural erosion risks (Masango, 2010).

New technologies have impacted Africa's oral arts, which transmit knowledge, values, and identity across generations. Critics argue that mass media and digital networks risk eroding living heritage through paradigm shifts from relational to individual learning and decentralisation of knowledge authority (Kaschula & Mostert, 2009). Interactive oral performance becomes passive media consumption. Customary protocols around sharing wisdom risk being ignored as content flows freely online, absent traditional safeguards (Alemna, 1998). However, new tools also enable innovative preservation and distribution of oral arts. Digital archives, online journals, and databases on indigenous knowledge democratise access to vanishing stories and techniques. Hyperlinked multimedia formats enhance contextualisation and accessibility. Media dissemination raises awareness of marginalised local cultures. Creative digital expressions like animations visualise ancestral narratives (Bulkani et al., 2022).

Synergies are possible when new platforms respectfully amplify living heritage as ongoing, not static. Online dimensions can augment physical learning, building demand for embodied practices. Collaborative digitisation enables elder guidance on what and how to share from customary knowledge banks. Platforms facilitating two-way dialogue and user contribution deepen engagement with living traditions (Hovik & Giannoumis, 2022). However, formats risk severing oral arts from ritual contexts and communal ethics vital for meaningful transmission. Imposing external logic of efficiency and monetisation for cultural extraction undermines traditional protocols for responsibly sharing wisdom. Frictions arise when new media displace, rather than enrich, living heritage and customary knowledge systems adapted to African contexts. Cultures undergoing rapid digitisation require strengthened resources and rights to self-determine appropriate integration pathways guided by, not disruptive to, African values and epistemologies (Mawere & Van Stam, 2019).

Evidence of technology's complex effects reveals both creative synergies and problematic disruptions. When performers innovatively adapt new tools to expand intergenerational transmission, expressive capacity, and access, technology becomes an avenue to sustain communal heritage. However, critics argue that digitisation risks eroding vernacular diversity and holism when embedded in paradigms of commodification, decontextualisation, and knowledge extraction (Barber, 2009). Positive examples include digital platforms enabling remote learners to connect with elderly storytellers in simulated apprenticeship and video documentation, helping sustain endangered story forms through accessible archives (Sehrawat et al., 2017). Creative projects like Bongani Sitole's hybrid poems fuse praise poetry with hip-hop, demonstrating a synergistic fusion of traditional aesthetics and sampling (Kaschula & Mostert, 2009). However, disruptive effects outweigh the benefits if digitisation facilitates the misappropriation or devaluation of living heritage as artefacts detached from communal ethics. Equipping youth "tradition-bearers" with media skills to creatively sustain oral arts demonstrates positive adaptation, as does carefully considered open sharing catalysing revitalisation. Conversely, uncritical adoption of formats like websites or e-books structured around external logic risks digitising the oral word in ways that disrupt its ritual essence and cultural functions. User-generated content can dilute accuracy and authority if detached from traditional protocols.

4.3   Impacts on Language and Communication

Modern digital technologies and global communications interfaces transform language practices across Africa's over 2000 diverse tongues. While presenting new opportunities, these rapid technolinguistic changes also pose risks requiring thoughtful navigation. Influence flows multidirectionally as major international languages shape technological systems, yet creative grassroots adaptation also sustains marginalised vernaculars online (Osborne, 2010). Dependence on Western digital platforms subtly imposes linguistic conventions privileging exogenous vocabularies over African idioms: keyboard limitations, voice recognition bias, and word processing tools model English structure, constraining tonal renderings (Djité, 2008). However, African youth are also ingeniously shaping technological spaces, as seen in multilingual social media connecting diasporas to home languages. Code-switching blends English, Swahili, dialects, and tech terms in dynamic speech, asserting identity (Wawire, 2017). Furthermore, vernacular hashtag campaigns celebrate linguistic heritage, while tech vocabulary is creatively vernacularised by youth through social media.

Nevertheless, online visibility favours primary cross-border tongues over hyperlocal ones. Marginalisation risks language endangerment as speakers shift usage to dominant languages underrepresented digitally.

Elders passing with endangered dialects take irreplaceable knowledge. However, promoting multilingual equity online is vital for sustaining identity and knowledge diversity. Policy interventions like standardising major African languages for educational and technical domains are debated, though risks neglecting unwritten vernaculars facing extinction (Maseko & Kaschula, 2009). Therefore, collaborative digital archives and multilingual interfaces help sustain these languages.

Creative linguistic hybridity emerges digitally as African youth fuse global hip-hop aesthetics with local lexicons. Tech spread does not simply homogenise but fosters new blends. Urban slang incorporates African orality, like proverbs with borrowed terms, into innovative speech acts. Transliteration renders oral phrases into written form online as youth bridge generations. Tech enables linguistic dynamism. These complex flows reveal both creative possibilities and problematic impacts at the intersection of African languages, technology, and culture. Challenges of displacement, commodification, and erosion persist due to unequal power structures. However, using available tools, vernacular communities also exercise agency in shaping inevitable change on their terms (Djité, 2008). Guiding language transitions requires recognising diversity, contingency, and grassroots innovation. Kneejerk technophobia is unproductive amid rapid change, as is an uncritical embrace of new media. While risks are real, collaborative technology integration on African terms holds promise to sustain the vitality and diversity of Africa's robust linguacultural heritage and knowledge systems in today's digitally interconnected world.

Digital technology and Internet connectivity have significantly influenced African languages, introducing novel writing, speaking, and vocabulary development modes. Critics argue that dependence on Western devices and platforms subjugates African linguistic heritage (Darvin, 2016). English conventions become embedded in word processing and social media structuring writing. Keyboards limit tonal renderings. Dictation software recognises primary tongues over unwritten vernaculars. Novel vocabularies also emerge around digital activities like 'scrolling', 'browsing', and 'tweeting' (Barber, 2009). As noted, urban youth develop slang by integrating African lexicons with global technoculture and engaging in code-switching, blending English, vernaculars, and tech terminology in dynamic speech acts. Such linguistic shifts catalysed by digital technology generate tensions between language preservation and cultural evolution.

Purists may argue that imported jargon erodes African lexicons and expressions optimised over generations to local experiences and that technocratic English dominates websites and policy documents due to limited digital content in vernaculars. Further, machine mistranslation propagates inaccuracies between languages, and technological determinism risks undermining idiomatic wisdom encoded in indigenous tongues (see Bulhan, 2015). However, integrative perspectives like moderate communitarianism recognise cultural dynamism and the potential for linguistic decolonisation through tactical code-switching to help assert African identity within global systems. Strategic hybridity blends technological functions with African lingua culture.

Power dynamics between languages are impacted as global tongues like English dominate digital spaces and technical domains. Marginalisation of indigenous languages risks language endangerment as speakers shift usage towards dominant externals (Makalela, 2016). Intergenerational transmission declines without digitised content, keyboards, and machine learning supporting vernaculars. Online visibility favours languages of wider communication. Only a small percentage of Africa's 2000+ languages are encoded into digital systems (Deumert & Lexander, 2013). Swahili, Yoruba, Hausa, and other widely spoken tongues accessed online gain prestige over smaller, unwritten vernaculars. Some advocate standardising major African languages for technical domains to foster self-reliance and negate dependence on former imperial tongues (see Mazrui & Mazrui, 1999). Knowledge digitisation projects like the African Languages Digital Library increase online content accessibility (Ngcobo, 2014). Localised software interfaces enable digital participation without linguistic assimilation. Hashtag campaigns celebrate language pride. However, most advocacy focuses on cross-border tongues more than hyperlocal unwritten vernaculars facing existential threat.

4.4   TECHNOLOGY, AND ARTISTIC EXPRESSION AND AESTHETICS

Technology has influenced African artistic expression and aesthetics, enabling new forms, materials, and distribution channels. The emergence of digital technology and the Internet has provided African artists with new tools and platforms to create, share, and promote their work (Bisschoff, 2017). This has led to the development of a vibrant digital art scene in Africa, where artists utilise digital tools and techniques to explore new forms of expression and engage with global audiences. The accessibility and affordability of digital tools have allowed artists to experiment with different mediums and techniques, expanding the possibilities of artistic creation (Zilberg & Steiner, 1996). However, integrating technology into artistic expression has created tensions between traditional arts and digital media. Some artists and communities may view digital tools as a departure from traditional artistic practices and aesthetics. There is a concern that the proliferation of digital media may overshadow or devalue traditional art forms, leading to a loss of cultural heritage and identity. This tension between tradition and innovation highlights the need for a balanced approach that respects and preserves traditional arts while embracing the opportunities offered by technology.

Furthermore, new tools expand possibilities for African artistic expression using novel techniques, materials, and distribution platforms. Photographic media enabled documentation of cultural practices and environments from colonisation, and news outlets and magazines fostered journalism, graphics, and satire at scale (see Vokes & Newbury, 2018). Cellphone cameras sparked citizen photojournalism and vernacular memes. Digital media facilitate the manipulation of images and sound for new effects. 3D printing crafts novel objects integrating traditional symbolism with emerging materials and forms. Online access has exponentially widened distribution, though often privileging Western aesthetics. All this has been celebrated for expanding global creative possibilities and access and visibility of African arts. Technology has become a means for cultural archiving and revival. However, others argue that digital mediums risk rupturing the living essence of oral heritage and ceremony expressed through ephemeral arts (Kaschula & Mostert, 2009). Pressure emerges to commodify art for Western consumption. Questions persist on guiding change processes under African terms. As such, uncritical adoption of foreign technologies risks eroding the communal values, spiritual dimensions, and situated knowledge systems that have long shaped African artistic expression. Decontextualised digitisation of traditional arts as static artefacts for global circulation can undermine the embodied, participatory, and sacred aspects of African aesthetics.

Social media and digital platforms have also shaped African artistic expression and aesthetics. These platforms have provided artists with new ways to connect with audiences, share their work, and collaborate. African artists have used social media to document and disseminate their creative processes, engage in online debates, and showcase their art to a global audience. This has allowed for the democratisation of artistic expression, as traditional gatekeepers no longer limit artists, who can now directly connect with their audience. The use of social media has also facilitated the formation of online communities and networks, enabling artists to collaborate and exchange ideas across geographical boundaries. However, rapid digitisation generates friction with traditional arts tuned to embodied presence, tactility, and living transmission. Critics warn that technology fosters imbalances as local arts are reformulated for external tourist demand and profit over sustaining communities (Grabski, 2017). In music, drumming and dance risk being reduced to digital samples appropriated without consent, and the loss of ritual spaces displaces arts from spiritual functions to decoration. Further, online dissemination may ignore customary knowledge-sharing protocols under elder guidance. However, from a moderate communitarian perspective, thoughtful integration of appropriate technology could enrich tradition with expanded expressive reach.

Moreover, technology has enabled African artists to explore new forms of artistic expression and experiment with different mediums. For example, digital storytelling has become a powerful tool for community engagement and social advocacy in Africa (see Treffry-Goatley et al., 2016). By combining oral storytelling traditions with computer and video production technology, digital storytelling allows communities to share their stories and perspectives in a compelling and accessible way. This has provided a platform for marginalised voices to be heard and has contributed to a more diverse and inclusive artistic landscape in Africa (Treffry-Goatley et al., 2016). Examples demonstrate technology carefully integrated to sustain, not displace, African innovation and aesthetics. Artists like Cyrus Kabiru creatively upcycle electronic waste into culturally resonant eyewear sculptures (Sebambo, 2015).

In some cases, technological tools have been successfully fused with cultural aesthetics, resulting in innovative and unique artistic expressions. Afrofuturism is one such example, where artists combine elements of science fiction, fantasy, and African aesthetics to imagine alternative futures for the African diaspora. Digital technology has allowed Afrofuturist artists to create immersive and interactive experiences that challenge traditional notions of art and storytelling. By incorporating digital tools and techniques, these artists can connect ancestral narratives of resiliency, hope, spirituality, and strength with a vision of the future that embraces

Black lives. Afrofuturism blends African motifs, fantasy, and mythology with sci-fi imaginaries as self-affirming speculative art (Peattie, 2021).

In a nutshell, technology has significantly impacted artistic expression and aesthetics in Africa. It has enabled new forms, materials, and distribution channels, providing artists with tools to explore innovative artistic expressions. However, technology integration has also created tensions between traditional arts and digital media. The fusion of technological tools and cultural aesthetics, as seen in Afrofuturism, showcases the potential for innovative and unique artistic expressions. Social media and digital platforms have further democratised artistic expression, allowing artists to connect with global audiences and collaborate with other artists. Thus, navigating the complex intersection of African arts, technology, and culture requires a balanced perspective that recognises risks and opportunities, fosters critical adaptation over passive assimilation, and privileges grassroots agency in shaping inevitable transformations.

4.5   SOCIAL INSTITUTIONS AND PRACTICES

Modern technologies impact African social institutions and communal practices, catalysing complex socio-cultural effects. Digital networks, automated systems, and urbanisation reshape sites of production, exchange, and authority. However, rapid transformations risk undermining enduring cultural logics of solidarity and dignity when change processes lack thoughtful foresight. Technologies enable opportunities like expanding civic networks, cooperation, and access to services through digital infrastructure. However, external technical systems oriented towards industrial efficiency frequently displace holistic human relationships and oral knowledge transmission vital to social cohesion. Unreflective adoption of foreign paradigms risks fragmentation by platforms optimising analytics over ethics. Creating collective African social fabrics requires navigating complex transitions with care and wisdom. Technologies must be adapted to augment communal spaces, not fracture them. However, reclaiming cultural agency to shape inevitable changes requires more than introducing novel artefacts. Policy reforms must nurture communal rights and foresight capabilities over purely technical installation. Education integrating African knowledge systems enables a savvy evaluation of socio-cultural impacts (Kaya & Seleti, 2013).

Technologies alter communal spaces, practices, and demographics in Africa as systems optimise industrial efficiency and global connectivity over locality. Urbanisation and digital mediation deterritorialised production, exchange, and authority from specific sacred places and older custodians towards dispersed factories, offices, and online platforms (Barber, 2006). Telework and digital shopping diminish collective ritual spaces. Automation changes agricultural roles and knowledge transmission. Youths who are proficient in new technologies see status gains over elder authority grounded in customary skills threatened by automation and displacement. Online anonymity and individualism erode communal accountability. Networked coordination alters community decision-making towards representational politics mediated at a distance or through social media. Face-to-face oral transmission shifts to digital platforms controlled by external regimes. Such modifications can enable inclusivity but risk fragmenting holistic social fabrics and spaces fundamental to African existence.

However, thoughtful adaptation can blend old and new to enrich African communal life. Mobile money services like M-Pesa expand financial inclusion by integrating with communal savings groups. Online forums connect diaspora communities to home villages. Telemedicine complements traditional healing through remote care. Virtual marketplaces sustain informal trade relationships. Social media amplifies grassroots organising around shared interests, and participatory mapping strengthens communal land rights advocacy (Hohenthal et al., 2017). Collaborative digital storytelling renews intergenerational knowledge sharing. Such hybrid approaches recognise technology as malleable to contextual needs when guided by African values of reciprocity, kinship, and restorative justice. Realising technology's potential to serve African communal fabrics requires proactive stewardship. As we will see later, policymakers must strengthen digital literacy, data sovereignty, and participatory governance to make decision-making more participatory. Africa must develop proper regulatory frameworks to protect collective intellectual property and indigenous knowledge from misappropriation. Furthermore, investments in community-based innovation ecosystems can incubate contextual solutions. Reconceptualising infrastructure beyond technical installation towards cultivating shared socio-cultural capabilities offers a holistic approach.

Influxes of foreign technologies calibrated for efficiency and global connectivity frequently erode established African social institutions and relationships. For instance, oral knowledge systems reliant on elder guidance risk being displaced by online media, fostering misinformation absent traditional accountability. Moreover, as with cartographic colonialism, external solutions risk undermining local expertise and governance (Harley, 1988). Similarly, modern individualism propagated through media and consumer technologies contradicts collectivist values and consensual decision-making. Consequently, networked coordination alters organic community relationships, accentuating discord without reconciliatory mechanisms (Atzori, 2015).

Furthermore, labour-saving devices disrupt reciprocal local economies, the density of interactions, and social solidarity when deployed without forethought. This cultural self-determination requires sustaining communities on African terms, not transitioning directly to the alien logic of atomisation and displacement. However, a kneejerk rejection of all technology risks forgoing benefits. Instead, more promising pathways involve thoughtful adaptation guided by ethical principles and African cosmologies to sustain collective social fabrics creatively. Ultimately, navigating the complex intersection of technology and African social institutions demands a nuanced approach that recognises risks and opportunities while privileging communal agency in shaping inevitable transformations.

At the same time, technology integration spawns novel institutions, forms of belonging, and values in Africa. For example, urban migration and digital lifestyles foster cosmopolitan identities, blending local and global influences. New civil society networks also expand civic participation beyond kin and locality (Adebanwi, 2011). Technological hubs spur entrepreneurial collaboration and peer learning, while online platforms create mediated communities, transcending physical barriers. However, such emergent techno-social formations reflect external paradigms dominating African worlds through purportedly neutral systems (Ayisi, 1992). Nevertheless, the reality is more complex. Novel institutions like Nollywood integrate African creative aesthetics with global dispersion, and social media sustains diaspora linguistic and cultural connections (see Bosch, 2017). Rather than a false binary between status quo tradition versus externalised modernity, space exists for new institutions and identities that synergise African social ethics and kinship with expanding tech-enabled networks on African terms.

4.6   RECOGNISING AFRICA'S DIVERSE
TECHNOCULTURAL IMPACTS

The rapid influx of modern technologies across Africa has catalysed diverse socio-cultural effects ranging from creative synergy to problematic disruption of local fabrics. However, linear development narratives assuming unidirectional cultural loss from technology insertion fail to capture complex on-the-ground realities. Instead, a more nuanced examination through an ethnographic lens reveals contingent, uneven, and negotiated adoption pathways dependent on contextual factors shaping communal agency and capabilities.

In sites where performers, activists, and artisans innovatively adapt technologies to expand intergenerational knowledge transmission and creative capacity on their terms, tools become avenues to sustain cultural evolution rooted in traditional ethics and practices. For instance, examples like amplifying marginalised languages through collaborative digital archives, leveraging mobile tools to expand youth civic engagement, and blending oral storytelling with digital media under elder guidance showcase thoughtful integration of sustaining heritage with expanded participatory reach (Bosch, 2017). However, the uncritical influx of external technologies optimised for industrial efficiency, standardised production, and global connectivity has also widely disrupted and displaced local institutions, knowledge systems, and intangible heritages in Africa over decades of uneven adoption. In particular, where change processes are driven by external interests rather than inward-looking communal needs and oversight, grave losses of living culture result. Moreover, even well-intentioned interventions may inadvertently undermine situated learning, reciprocal economies, and social cohesion by inserting decontextualised technical models (Brynjolfsson & McAfee, 2014).

These diverse narratives of creative adaptation and problematic disruption complicate linear cultural evolution or erosion assumptions. The impacts of any technology insertion are shaped by contextual factors mediating community agency, resources, demographics, policies, and capabilities to direct or resist change. Furthermore, place, people, and process contingencies foster divergent indigenisation pathways and socio-cultural effects. Therefore, responsible innovation requires a granular analysis of heterogeneous African realities and diverse human needs, driving bottom-up change rather than importing one-size-fits-all technical solutions (Mawere & Van Stam, 2019). Mindfully bridging old and new through participatory design grounded in African cosmologies holds the potential to sustain cultural dynamism and self-determination. However, this necessitates recognising communities as agents, not just consumers of technology. Moreover, understanding contextual complexities and indigenous ethics is vital to shift paradigms from disruptive modernisation towards thoughtful localisation that protects shared heritages while enabling continuity (Mawere & Awuah-Nyamekye, 2015).

The influx of novel tools has fostered user-led innovation, adapting technologies to dynamic local contexts across urban and rural zones. For instance, youth culture blends global hip-hop aesthetics with African lingua culture, identities, and instrumentation to sustain creative dynamism in African terms. Additionally, public libraries leverage digital archiving to preserve and share local knowledge threatened by urbanisation (Okocha, 2022). Moreover, telehealth networks creatively expand medical access and community healthcare information by interweaving digital systems with communal support structures (Haleem et al., 2021). Further, hackerspaces foster peer learning and digital fabrication with locally available materials, needs, and ingenuity as inputs rather than mimicry of foreign models (Fox et al., 2015). Similarly, agricultural cooperatives integrate mobile finance platforms and data tools to scale inclusive, ethical business models rooted in reciprocity. By synergising African social philosophies with connectivity, technology becomes a conduit for grassroots socioeconomic advancement, not disruption.

The uncritical influx of external technologies optimised for industrial efficiency, standardised production, and global paradigms has also widely eroded local practices, institutions, and heritage. For example, biopiracy removes indigenous medicinal knowledge from communal control through exploitative patenting and commodification (Mgbeoji, 2006). Similarly, imported monocultures displace resilient seed diversity and agroecological farming techniques intergenerationally adapted to African soils and climates, undermining food sovereignty. Moreover, the insensitive digitisation of heritage for Western museums, from objects like Benin bronzes to botanical specimens, extracts artefacts as property detached from living cultures (Brah & Coombes, 2000). Such decontextualised appropriation contrasts with collaborative digitisation, amplifying marginalised local voices and knowledge.

These complex narratives reveal the potential and pitfalls of technology's intersection with African socio-cultural fabrics. Thoughtful integration guided by communal ethics and epistemologies can catalyse creative continuity and expanded capacities. However, uncritical adoption driven by external interests frequently erodes situated knowledge, institutions, and heritages. Therefore, responsible innovation demands participatory processes centring on African agencies, realities, and value systems to democratically shape socio-technical change (Mawere & Van Stam, 2019). As argued earlier, regulatory frameworks protecting collective intellectual property, data sovereignty, and cultural rights are vital. Additionally, investments in community-led innovation ecosystems, context-appropriate infrastructure, and endogenous knowledge integration within education can strengthen adaptive capabilities.

The unreflective importation of Western education models reliant on standardised curricula and digitally mediated content risks displacing situated, embodied learning and knowledge transmission that are vital for the socialisation of youth on African terms. Moreover, public media systems and consumer technologies propagating individualism erode oral traditions, consensual governance, and collectivist values (Chabal, 2009). Similarly, automation alters traditional reciprocal economies without considering equity impacts. Where change processes are driven violently by external interests rather than community needs, grave loss of intangible and embodied heritages ensues. However, some disruptions may inadvertently arise from well-intentioned interventions that are ignorant of local contexts. Therefore, responsible innovation requires recognising the diversity of situated socio-cultural realities shaping technology integration pathways and effects (Smyth et al., 2010).

Varying responses across Africa's heterogeneous environments caution against deterministic assumptions of the necessary sacrifice of traditional forms for modernisation. For instance, rural villages sustain reciprocal economies and oral rituals even amid partial technology adoption, enabled by demographics and resistant agency (Nyamnjoh, 2012). Urban youth readily remix foreign media content and digital networks with local aesthetics to define new creative identities. Moreover, women's informal networks leverage mobile phones to sustain collective solidarity and livelihoods, partly shielding against individualising impacts (Kabeer, 2019). Furthermore, communities retaining democratic oversight of infrastructure and resources can better manage disruptions through participatory technology assessment and foresight. Contextual specificity shapes technological integration and effects. Consequently, responsible innovation requires careful analysis of diverse situated realities and contingency rather than importing one-size-fits-all technical solutions.

The binary juxtapositions of timeless tradition versus modernity are not plausible. Instead, space exists for thoughtful synthesis if change processes are guided by communal principles of continuity, sustainability, and social justice marginally obscured within conventional technoscience discourses. Moreover, mindfully directed technological transformation following African ethical perspectives holds the potential to catalyse inclusive cultural evolution and creativity anchored in living heritage. This requires centring African philosophies, values, and knowledge systems as the foundation for collaborative socio-technical design and governance. Furthermore, participatory processes engaging diverse community stakeholders, from elders to youth, can surface contextual needs, priorities, and cultural sensitivities to inform appropriate innovation pathways.

Investments in endogenous innovation ecosystems, community-led infrastructure, and culturally responsive education integrating African epistemologies can strengthen adaptive capabilities. Moreover, regulatory frameworks protecting communal intellectual property, data sovereignty, and collective cultural rights are vital to mitigate exploitative appropriation (Oguamanam, 2018). Harmonising African social fabrics with technological change ultimately demands privileging communal self-determination, epistemic justice, and holistic well-being as guiding innovation principles. Through such decolonial processes, Africa's civilisational wisdom can inform the repurposing of socio-technical systems to advance Ubuntu in the digital age while creatively sustaining the shared vitality of tradition and innovation.

4.7   Conclusion

This chapter has examined diverse evidence on technology's complex intersections with dimensions of African culture and argued against linear assumptions of necessary displacement of traditional forms through modern technologies. While digital tools and external paradigms risk disrupting embodied arts, languages, social fabrics, and values, thoughtful adaptation also reveals possibilities to sustain cultural dynamism. However, neglecting African contexts and agency in prevailing innovation ecosystems poses dangers of erosion. Guiding appropriate integration pathways necessitates centring African conceptual resources and communal principles marginalised within conventional technoscience discourses. This concluding analysis synthesises key lessons and proposes reform directions to strengthen cultural self-determination through mindful localisation.

These complex, contingent processes demand nuanced assessment of varied socio-cultural effects to inform policy reforms. Responsible innovation requires strengthening communal capabilities and rights to direct change, not just introducing new artefacts and infrastructures devoid of context. Technology aligned with African philosophies and engineered through participatory design can enrich cultural dynamism and collective advancement if anchored in continuity with wisdom traditions. This chapter has highlighted the dangers of narrow technocratic paradigms that treat culture as static artefacts rather than living heritages. However, possibilities emerge by recognising Africa's diverse cosmologies, ethics, and ways of knowing as invaluable epistemic resources to guide appropriate technological integration.

Beyond just enabling modernisation, thoughtfully localised technologies can catalyse innovative expressions of identity and knowledge transmission rooted in community needs and values. Fulfilling these possibilities necessitates elevating African voices to reshape prevailing innovation ecosystems that obscure local agencies. Policies must nurture participatory models of technology development and diffusion responsive to situated contexts. Education and research require integrating African knowledge traditions and design thinking marginalised under conventional technoscience. Technology can only become a tool for cultural flourishing rather than disruption by re-centring control and agency within Africa's diverse communities.

Overall, navigating the transformations catalysed by accelerating technological change across Africa demands a fundamental reorientation of innovation processes and value systems. Centring African philosophies, participatory design, and communal governance can guide the collective repurposing of socio-technical systems to advance Ubuntu and holistic well-being sustainably. By critically adapting tools through collaborative localisation, the wisdom of Africa's civilisational heritage offers generative ethical resources to harmonise tradition and innovation in our digitally entangled world. Through such decolonial approaches, technology can become an empowering conduit for Africa's diverse communities to shape cultural evolution inclusively and creatively on their terms.

REFERENCES

Adebanwi, W. (2011). The clergy, culture, and political conflicts in Nigeria. *African Studies Review, 54*(3), 88–87.

Akinwale, A. A. (2012). Digitisation of indigenous knowledge for natural resources management in Africa. *Environmental Science, Sociology.*

Alemna, A. A. (1998). An overview of library and information research in West Africa. *African Journal of Library, Archives and Information Science, 8*(1), 1–12.

Asuquo, M. E., Okenjom, G. P., Okpa, O. E., & Eyiene, A. E. (2023). Opportunities and challenges in digitization of indigenous knowledge and implication for educational management in the Nigerian Context. In T. Masenya (Ed.), *Digital preservation and documentation of global indigenous knowledge systems* (pp. 87–104). IGI Global. https://doi.org/10.4018/978-1-6684-7024-4.ch005

Atzori, M. (2015). Blockchain technology and decentralised governance: Is the state still necessary? *SSRN Electronic Journal.* https://doi.org/10.2139/ssrn.2709713

Ayisi, E. O. (1992). *An introduction to the study of African culture.* East African Publishers.

Baker, C. (2012). The evolution of language policy in Africa. In *The Cambridge handbook of language policy* (pp. 275–292). Cambridge University Press.

Barber, K. (2006). *Africa's hidden histories: Everyday literacy and making the self.* Indiana University Press.

Barber, K. (2009). Orality, the media and new popular cultures in Africa. In K. Njogu & J. Middleton (Eds.), *Media and identity in Africa* (pp. 3–18). Edinburgh University Press.

Bisschoff, L. (2017). The future is digital: An introduction to African digital arts. *Critical African Studies, 9*(3), 261–267. https://doi.org/10.1080/21681392.2017.1376506

Bosch, T. (2017). Twitter activism and youth in South Africa: The case of #RhodesMustFall. *Information, Communication & Society, 19*(2), 221–232.

Brah, A., & Coombes, A. E. (2000). Introduction: The conundrum of 'mixing'. In *Hybridity and its discontents* (pp. 1–16). Routledge.

Brynjolfsson, E., & McAfee, A. (2014). *The second machine age: Work, progress, and prosperity in a time of brilliant technologies.* W.W. Norton & Company.

Bulhan, H. A. (2015). Stages of colonialism in Africa: From occupation of land to occupation of being. *Journal of Social and Political Psychology, 3*(1), 239–256. https://doi.org/10.5964/jspp.v3i1.143

Bulkani, Fatchurahman, M., Adella, H., & Setiawan, M. A. (2022). Development of animation learning media based on local wisdom to improve student learning outcomes in elementary schools. *International Journal of Instruction, 15*(1), 55–72. https://doi.org/10.29333/iji.2022.1514a

Chabal, P. (2009). *Africa: The politics of suffering and smiling.* Zed Books.

Darvin, R. (2016). Language and identity in the digital age. In S. Preece (Ed.), *Routledge handbook of language and identity.* Routledge.

Deumert, A., & Lexander, K. V. (2013). Texting Africa: Writing as performance. *Journal of SocioLinguistics, 17*(4), 522–546.

Djité, P. G. (2008). *The sociolinguistics of development in Africa.* Multilingual Matters.

Finnegan, R. (2012). *Oral literature in Africa.* Open Book Publishers.

Fox, S., Ulgado, R. R., & Rosner, D. (2015, February). Hacking culture, not devices: Access and recognition in feminist hackerspaces. In *Proceedings of the 18th ACM conference on computer supported cooperative work & social computing* (pp. 56–68). Association for Computing Machinery

Grabski, J. (2017). *Art world city: The creative economy of artists and urban life in Dakar.* Indiana University Press.

Haleem, A., Javaid, M., Singh, R. P., & Suman, R. (2021). Telemedicine for healthcare: Capabilities, features, barriers, and applications. *Sensors International,* p. 2, 100117. https://doi.org/10.1016/j.sintl.2021.100117

Harley, J. B. (1988). Silences and secrecy: The hidden agenda of cartography in early modern Europe. *Imago Mundi, 40,* 57–76. http://www.jstor.org/stable/1151014

Hohenthal, J., Minoia, P., & Pellikka, P. (2017). Mapping meaning: Critical cartographies for participatory water management in Taita Hills, Kenya. *The Professional Geographer, 69*(3), 383–395.

Hovik, S., & Giannoumis, G. A. (2022). Linkages between citizen participation, digital technology, and urban development. In S. Hovik, G. A. Giannoumis, K. Reichborn-Kjennerud, J. M. Ruano, I. McShane, & S. Legard (Eds.), *Citizen participation in the information society.* Palgrave Macmillan. https://doi.org/10.1007/978-3-030-99940-7_1

Jewsiewicki, B. (1989). African historical studies: Academic knowledge as 'usable past' and radical scholarship. *African Studies Review, 32*(3), 1–76.

Kabeer, N. (2019). Randomised control trials and qualitative evaluations of a multifaceted programme for women in extreme poverty: Empirical findings and methodological reflections. *Journal of Human Development and Capabilities, 20*(2), 197–217.

Kaschula, R. H. (2004). Imbongi to slam: The emergence of a technologised Auriture. *Southern African Journal of Folklore Studies, 14*(2), 45–58.

Kaschula, R. H., & Mostert, M. L. (2009). Analysing, digitising and technologising the oral word: The case of Bongani Sitole. *Journal of African Cultural Studies, 21*(2), 159–175.

Kaya, H. O., & Seleti, Y. N. (2013). African indigenous knowledge systems and relevance of higher education in South Africa. *The International Education Journal: Comparative Perspectives, 12*(1), 30–44.

Makalela, L. (2016). Ubuntu translanguaging: An alternative framework for complex multilingual encounters. *Southern African Linguistics and Applied Language Studies, 34*(3), 187–196.

Masango, C. A. (2010). Indigenous traditional knowledge protection: Prospects in South Africa's intellectual property framework? *South African Journal of Libraries and Information Science, 76*(1), 74–80.

Maseko, P., & Kaschula, R. H. (2009). Vocational language learning and teaching at a South African university: Preparing professionals for multilingual contexts. *Stellenbosch Papers in Linguistics PLUS, 38*, 130–142.

Mawere, M., & Awuah-Nyamekye, S. (Eds.). (2015). *Between rhetoric and reality: The state and use of indigenous knowledge in post-colonial Africa.* Langaa RPCIG. https://doi.org/10.2307/j.ctvh9vwc4

Mawere, M., & Mubaya, T. R. (2016). *African philosophy and thought systems: A search for a culture and philosophy of belonging.* Langaa RPCIG. https://doi.org/10.2307/j.ctvk3gkz5

Mawere, M., & Van Stam, G. (2019). Research in Africa for Africa? Probing the effect and credibility of research done by foreigners in Africa. In P. Nielsen & H. C. Kimaro (Eds.), *Information and communication technologies for development. Strengthening Southern-driven cooperation as a catalyst for ICT4D* (pp. 168–179). Springer.

Mazrui, A. A., & Mazrui, A. M. (1999). *Political culture of language: Swahili, society and the state.* Africa World Press.

Mgbeoji, I. (2006). *Global biopiracy: Patents, plants, and indigenous knowledge.* UBC Press.

Ngcobo, S. (2014). The struggle to maintain identity in higher education among Zulu-speaking students. *International Journal of Bilingual Education and Bilingualism, 17*(6), 695–713.

Ngulube, P. (2002). Managing and preserving indigenous knowledge in the knowledge management era: Challenges and opportunities for information professionals. *Information Development, 18*(2), 95–102.

Nyamnjoh, F. B. (2012). 'Potted plants in greenhouses': A critical reflection on the resilience of colonial education in Africa. *Journal of Asian and African Studies, 47*(2), 129–154.

Oguamanam, C. (2018). *Tiered or differentiated approach to traditional knowledge and traditional cultural expressions: The evolution of a concept* (Ottawa Faculty of Law Working Paper No. 2018-28).

Okocha, F. (2022). Digital libraries in Africa: Challenges and opportunities. *Library Philosophy and Practice (e-journal).* 7288.

Osborne, D. (2010). *African languages in a digital age: Challenges and opportunities for indigenous language computing.* HSRC Press.

Owusu-Ansah, F. E., & Mji, G. (2013). African indigenous knowledge and research. *African Journal of Disability, 2*(1), 1–5.

Peattie, P. (2021). Afrofuturism revelation and revolution; Voices of the digital generation. *Journal of Communication Inquiry* https://doi.org/10.1177/01968599211041117.

Sebambo, K. (2015). *Kenyan artist creates eccentric eyeglasses from electronic waste.* Digital Indaba.

Sehrawat, S., Jones, C. A., Orlando, J., Bowers, T., & Rubins, A. (2017). Digital storytelling: A tool for social connectedness. *Geron, 16*(1), 56–61. https://doi.org/10.4017/gt.2017.16.1.006.00

Smyth, T. N., Etherton, J., & Best, M. L. (2010). MOSES: Exploring new ground in media and post-conflict reconciliation. In *Proceedings of the SIGCHI conference on human factors in computing systems* (pp. 1059–1068). ACM.

Taylor, D. (2016). *Performance.* Duke University Press.

Treffry-Goatley, A., Lessells, R., Sykes, P., Bärnighausen, T., de Oliveira, T., Moletsane, R., & Seeley, J. (2016). Understanding specific contexts of antiretroviral therapy adherence in rural South Africa: A thematic analysis of digital stories from a community with high HIV prevalence. *PLoS One, 11*(2), e0148801.

Vokes, R., & Newbury, D. (2018). Photography and African futures. *Visual Studies, 33*(1), 1–10. https://doi.org/10.1080/1472586X.2018.1424988

Wawire, G. N. (2017). English-Swahili code switching: An intersection between stance and identity. *Electronic Theses and Dissertations.* 1238. https://egrove.olemiss.edu/etd/1238

Zeleza, P. T. (1997). *Manufacturing African studies and crises.* CODESRIA Books Publication System.

Zilberg, J., & Steiner, C. B. (1996). *African art and the colonial encounter.* Indiana University Press.

CHAPTER 5

Rethinking Technology Through African Philosophy

5.1   INTRODUCTION

Modern technologies like digital platforms, biotech, and AI are rapidly spreading across Africa. Mainstream Western ideas view technology as value-neutral tools for control, efficiency, and dominating nature. However, these paradigms ignore social impacts and cultural contexts. Alternative frameworks centred on human dignity and collective advancement are needed in Africa. Clarifying empowering, appropriate conceptions of innovation rooted in African thought is vital. This chapter contrasts Western and African perspectives on technology through philosophical analysis. If imposed uncritically, new technologies risk exacerbating inequalities and disrupting the social fabric in Africa. Core African philosophies articulating communal values, sacred ecology, and humanism offer alternative foundations. They conceptualise technology as instruments that enable or undermine collective well-being, social harmony, and human dignity. Whereas Western notions see technologies as value-neutral tools for control, African perspectives emphasise interdependence, care ethics, and co-existence between technology and human development. African philosophies provide different bases for conceptualising and employing technologies across sectors in empowering, socially responsible ways.

This chapter attempts to construct an African technology philosophy guided by principles of community, sacred ecology, and care rather than

Y. Ndasauka, *African Mind, Culture, and Technology*, https://doi.org/10.1007/978-3-031-62979-2_5

technical rationalism. An Afrocentric framework envisions technologies designed to harmonise with, not dominate, nature and society. Participatory, decentralised innovation models expressing African communal values and identities become building blocks to democratise and localise technological design. After outlining dominant Western conceptions and proposed alternatives, the chapter extracts pragmatic principles from African thought to guide shaping technology focused on holistic human development rather than just accumulation or control. This interrogation of prevailing paradigms and articulation of more empowering African frameworks aims to catalyse discourse on reforming policies, education, design, and governance. The chapter concludes by proposing African perspectives as vital for transforming technological thinking to align innovation with care, dignity, and social justice. The goal is to move beyond instrumental views of technology towards an African philosophy envisioning technology designed for communal well-being, social harmony, and human dignity.

5.2   WESTERN CONCEPTIONS OF TECHNOLOGY

Western philosophy has long seen technology as a value-neutral tool to control nature. Francis Bacon promoted using science to methodically uncover nature's secrets through experimentation alone to 'bind her to your service' (Bacon, 1920). This views technology as an instrument to subdue external nature, linked to modernity's project of freeing humans from biological existence. Tool use set enlightened humans apart from animals, enabling apparent environmental mastery and self-determination. In this worldview, technology becomes a way to curb natural forces, maximise productivity, and impose order on wilderness, chance, and limits. Technology was not seen as value-laden but as a neutral tool for human ends. Mainstream technical rationality aimed to optimise efficiency, productivity, and accumulation through technological control.

The Scientific Revolution promoted mechanical views of living things as passive, unstoppable machines that could be manipulated for human benefit through strict technical control and efficiency. Natural phenomena were seen as raw materials that could be exploited for material progress rather than entities with their sacredness or rights. Experimental science and quantification promise to uncover nature's hidden laws, allowing prediction and mastery through technology. In these narratives, technological progress matched scientifically optimising control, standardisation, and instrumental power over perceived natural randomness. The hypothetical ability to perfectly predict and control externals through technical means seemed to minimise human dependence on nature. However, critics argue that this paradigm of technology as mastery over nature risks imbalance (Ferré, 1988; Lemmens et al., 2017). Unrestrained instrumentalisation of the natural world endangers the sustainability of irreplaceable living systems humanity relies on despite illusions of independence. Furthermore, arrogance around passive natural objects ready for human control overlooks complex dynamics and inherent worth beyond human needs.

Modernity linked technology with means for liberated citizens to escape the supposed irrationality of traditions and hierarchies. Technology allowed enterprising individuals to exercise personal reason, creativity, self-reliance, and ingenuity in pursuits of enrichment, knowledge, and social mobility (Ferré, 1988). The scientific method and quantification expanded private, individualistic intellectual pursuits empowered by technologies like the printing press (De Vries, 1997). Universal material abundance through technological mastery became the promise of modernity. However, this notion of technologies as value-neutral tools autonomous individuals use risks atomisation and alienation. It overlooks how technologies shape cultural meanings and social relations in complex ways that cannot be predicted or controlled. Technologies are not simply neutral tools applied by individuals but exist within broader social contexts that influence their development and uses. While modernity promoted technology as a means of individual empowerment and freedom, technologies can also enable new forms of social control and conformity. The emphasis on scientific progress and technological innovation privileged some groups over others. Not all have equal access to emerging technologies or the ability to determine technical agendas. Rather than unambiguous liberation, technology holds ambivalent potential to both expand and constrain human capabilities and relationships.

Influential liberal thinkers like Descartes and Locke promoted views of the external world and untamed nature as passive objects open to exploitation and control by active, technically empowered individuals (Hanks, 2010). The self-determining individual could supposedly achieve domination over less enlightened others and nature through rational calculation and technological mastery. Competitive markets fuelled atomistic conceptions of self-interest maximisation enabled by continuous personal specialisation and technical efficiency. Technological power became synonymous with unconstrained individual freedom, advancement, and fulfilment within these common narratives. However, critics argue that such technology philosophies guided solely by aggregating individual gains without communitarian ethics may worsen inequalities, hierarchies, marginalisation, and fraying of social bonds. An unquestioned over-emphasis on personal technical mastery versus collective human improvement reflects unsustainable atomistic assumptions at odds with shared dignity (Feenberg, 1995).

Moreover, dominant Western technological paradigms promote continuous progress via more efficiency, productivity, predictability, rational order, and precise control in organising natural and human systems (Drengson, 1982). Scientific management techniques sought to optimise efficiency and normalise uncertainties through measurement, monitoring, standardisation, and control of production processes. Mechanisation, automation, and algorithmic decision-making aimed to reduce human manual effort and increase technical control through consistent, predictable, and efficient procedures. Digital technologies extend aspirations for automatic information control by potentially encoding complex environments into efficient, predictable machine processes (Lettow, 2011). In this narrative, advancing efficiency and technical supervision over nature and society through modelling and technology represents progress. Technocratic elites fluent in specialised engineering and statistical knowledge are tasked with optimising systems performance.

Feenberg (1991) highlights the dangers of an uncontrolled quest for technological control and efficiency without ethical guidance. Standardisation and imposed technical rules often oversimplify and disrupt organic social or ecological processes not reducible to mechanics. Biases and dehumanising commodification embedded in automated systems managing human affairs pose risks (Feenberg, 1991). High-modernist schemes seeking rational order have often damaged societies and environments, from urban slum clearances to authoritarian technocratic governance (Gunnell, 1982). These views propose democratically shaping technologies through public discourse to uphold pluralistic values rather than technical elites imposing scientific management. Community-centred innovation would enrich socio-ecological fabrics, not just maximise efficiency. This requires reconstructing technocratic mentalities that reduce nature and society to objects of control. An unguided embrace of technology risks disrupting organic systems and enables authoritarian control. Critics advocate inclusive public participation to uphold ethical values against technical elites imposing biased systems. Designing technology to enrich communities demands reforming reductionist technocratic mindsets.

According to Feenberg (1995), pursuing technological efficiency and control as ends-in-themselves discourages examining why particular optimisation goals are selected and for whose ultimate benefit. Unquestioned reliance on technical means and narrow performance metrics risks overlooking holistic social and environmental impacts. When efficiency becomes an unquestioned supreme value directing innovation, the result can be dangerously misaligned systems that worsen harmful externalities. Cost-benefit calculations aimed at efficiency maximisation absolve responsibility for social justice or sustainability (Winner, 1997). Alternative philosophies argue for balancing efficient functionality with ethical analyses of how technology reshapes relationships between people and nature. Purely quantifying progress via technical metrics risks losing sight of humanistic ends and pursuits. Rather than solely focusing on increasing efficiency and control, innovation guided by care, dignity and justice would examine localised needs, social impacts, and moral implications of optimisation schemes. The aim becomes appropriate, sustainable technology cooperatively governed for equitable benefit rather than impersonal systems maximising control (Mitcham, 1985).

Furthermore, mainstream Western paradigms of technology are reductionistically premised on ideals of controlling nature, maximising individual power, and pursuing efficiency, largely to exclude substantive ethical analysis, moral values, and considerations of collective welfare. This dominant paradigm assumes that technological design and governance should be guided solely by scientific facts, instrumental functionality, and value-neutral technical rationality. This risks marginalising normative concerns with humanistic social values and holistic development (Fallman, 2011). Positivist notions portray technological advancement as an autonomous, politically neutral scientific discovery and rational optimisation process. Nevertheless, scholars increasingly challenge these assumptions as ignoring morality, justice and technology's concrete impacts on human dignity and social relationships (Gardner, 1997). Beyond just functionality, technology inherently mediates social and political dynamics between differently empowered groups. Innovation guided by inclusive deliberation, not just technical control, may encourage more just distributions of benefits and burdens from modernisation (Gunnell, 1982). Moreover, excessive focus on quantifiable metrics risks commodifying human activities and nature while discounting meaningful pursuits exceeding technical measurement.

Critics propose re-centring substantive values discussions within technological thinking to complement functionality and efficiency considerations. This entails deliberate analyses of how technical systems concretely transform socio-ecological relationships and human capabilities in complex contexts (Feenberg, 2002). Innovation guided by care, dignity, and justice would examine needs, impacts, and moral implications rather than just control and metrics. Sustainable, cooperative technology requires integrating economic, social, and environmental objectives holistically. However, dominant intellectual traditions have largely excluded such humanistic analyses in deference to supposed value-neutrality. Technical rationality's unquestioned supremacy discourages examining underlying assumptions, exclusions, interests, and alternative possibilities embodied in innovation pathways. The quest for efficiency, productivity, and control risks entrenching exploitative relationships without addressing systemic injustices, inequalities, or environmental harms. As such, paradigm shifts recognising technology's normative dimensions become imperative.

When technology governance becomes excessively focused on narrow metrics of functionality and efficiency, broader consequences for marginalised populations are readily overlooked. Without proactive ethical guidance, technologies risk exacerbating injustice, unemployment, and socioeconomic disparities despite functioning optimally on paper. Technocratic design ideologies further obscure technology's socially constructed nature by naturalising particular configurations of power and expertise. Prevailing innovation patterns embedding oppressive biases are wrongly treated as inevitable rather than contested outcomes of malleable choices (Brey, 2010). Without deliberately centring substantive values, technologies guided solely by scientific technique risk tendencies towards dehumanisation and domination even when materially productive (Marcuse, 1964).

As discussed, re-centring humanistic social ethics, care, and justice within technology research, governance, and design is imperative. Innovation guided by ethical empowerment of marginalised peoples can counter tendencies towards domination by creating more participatory futures focused on universal human dignity and capabilities. However, this requires moving beyond metrics of functionality and efficiency. Integrating cultural meanings, inequality, and moral values into technology assessment and policy is necessary. Rather than assume technical means justify all ends, innovation-oriented cooperative human development requires guidance by moral philosophy from the outset. However, shifts are needed to overcome technology's pernicious exclusion of social values and ethics. Intellectual cultures resistant to such transformation dismiss moral concerns as secondary threats to scientific progress and functionality. Technocratic ideologies propagate myths of value-neutrality that obscure technology's mediation of complex values. Remedying this requires recognising that innovation is fundamentally normative, brokering difficult trade-offs (Brey, 2010). What is needed are designers, policymakers, and the public to be willing to shape technologies guided by justice and wisdom, not just control. Rather than defer to supposed neutral experts, participatory technology assessment can democratise defining values embedded in systems. Integrating multiple perspectives through deliberation, not technocratic imposition, is vital for upholding pluralism (Gunnell, 1982).

5.3   African Philosophies and Values

African philosophies offer alternatives to dominant Western worldviews focused heavily on individualism, self-interest, and technological domination over nature. African thought systems emphasise community interdependence, nature's sanctity, and care ethics beyond material or technical goals. In contrast to hyper-individualist Western ideas, African worldviews see human identity, status, and fulfilment as fundamentally rooted in reciprocity, social values and cooperation. As expressed in Ubuntu ideas across southern Africa, personhood is defined through mutual caring and 'I am because we are' (Mbiti, 1969). Rights and duties are based on contributing to communal well-being, countering Western liberal assumptions of self-interested individuals detached from social obligations. Equally, African cosmologies conceive human life as embedded within and dependent upon the sacred integrity of the natural order. Sustenance comes from respectful co-existence with other beings and environments. Humanity belongs to nature rather than apart from it. This sacred ecology diverges sharply from Western paradigms, seeing nature as a resource for human exploitation and domination through technology. African humanism prioritises care, dignity, and harmony beyond material or technical utility. Ubuntu values human individuals unconditionally through reciprocity, empathy, and contribution to the community. Peaceful conflict resolution matters more than punishment. These care ethics contrast with Western legalistic rules and atomised self-advancement.

In essence, African worldviews articulate different foundations from Western individualism, materialism, and technocracy. Rediscovering African thought marginalised under colonialism remains vital for countering anti-humanistic outcomes of market fundamentalism and techno-scientific control today. African communal values and dignity offer indispensable guidance for shaping more just, caring, and sustainable technological futures benefiting all humanity. However, prevalent technology paradigms struggle to integrate these different premises. Technocratic governance focuses on quantitative metrics, functionality and efficiency rather than community well-being, sacred nature or care ethics. Pursuing value-neutral technological control and productivity for individual gain conflicts with African principles of collective advancement through reciprocity. Hyper-individualist ideologies driving competitive markets, private property and self-interest are alien to philosophies defining personhood through social contribution.

Modernisation schemes seek to rationally order societies and dominate nature using scientific tools—directly countering sacral ecologies. Africa's philosophical alternatives disappear behind assumptions that Western paradigms signify universal progress. However, recovering African thought traditions marginalised under colonialism is indispensable for counteracting anti-humanistic tendencies of contemporary technology and economy. The urgent imperative is re-centring community, dignity, justice, ecology, and care at the heart of innovation and development. African cosmologies offer philosophical alternatives to dominant Western worldviews premised on radical individualism, atomistic self-interest, and ideals of technological domination over nature. Distinct humanistic foundations central to African thought systems emphasise communal interdependence, integrity of nature, and ethics of care beyond material or technical utility. In contrast to hyper-individualist Western conceptions, African philosophies recognise human identity, status, and fulfilment as fundamentally rooted in reciprocal social values and cooperative community relations. As expressed through the southern African principles of Ubuntu, personhood is defined through mutual caring. Rights and responsibilities are accorded based on contributing to communal well-being. This communitarian standpoint counters pervasive Western liberal assumptions of the self-interested individual abstracted from social obligations.

Equally, African worldviews conceive human life as embedded within and dependent upon the sacred integrity of the natural order. Sustenance derives from respectful co-existence with other living beings and environments. Humanity belongs to nature, not apart from it. This sacral ecology diverges sharply from Western paradigms of nature as a repository of resources for human exploitation and domination through technology. African humanism also centres ethics of care, dignity, and social harmony beyond material or technical utility. The concept of Ubuntu embodies valuing human individuals unconditionally through reciprocity, empathy, and contribution to the community. Peaceful dispute resolution matters more than punishment. These care ethics contrast with Western legalistic rules and atomised self-advancement. African philosophies thus articulate different foundations from Western individualism, materialism, and technocracy. Rediscovering African thought traditions marginalised under colonialism remains vital for countering the anti-humanistic outcomes of market fundamentalism and techno-scientific domination. The alternatives centred on African communal values and dignity offer indispensable guidance for shaping more just, sustainable and caring technological futures benefiting all humanity.

In the African tradition, human identity is fundamentally shaped through community ties and relationships rather than as an isolated, abstract individual (Menkiti, 1984). Unlike the Western liberal autonomous self, the African subject is social, gaining purpose and agency by enacting communal roles and contributing to shared ends. As Ubuntu principles express, an individual's status comes from displaying virtues benefiting the collective: 'A person is a person through other people' (Mbiti, 1969). Identity relies on empathetic alignment with the greater community. However, African communitarianism has been critiqued for overly conditioning individuality and dissent through conforming to dominant roles and hierarchies (Matolino, 2009). Moderate strands try to balance individual talents and communal values (Famakinwa, 2010). However, the enduring collective self remains distinct from the Western atomic individual. African communitarian ethics locate moral value in furthering the communal good rather than just personal benefit (Gyekye, 1997). Shared access to resources and opportunities signifies justice. Leaders carry duties as community stewards, not authoritarians. Dispute resolution prioritises social harmony over punishment. This can constrain liberties by subsuming individuals within totalising social claims. However, African thought recognises dissent, debate and dignity to balance obligations (Wiredu, 1980). People have duties alongside rights, preventing conformist oppression in solidarity's name. Some propose integrating African communal principles with egalitarian Western ideas to inspire social democracy through consensus (Van Niekerk, 2007). The communitarian grounding counters market fundamentalism's rupturing of social bonds.

In African customary law, violations destabilise the entire community, not just individuals. Communitarian justice focuses on restorative reconciliation to heal social relationships through inclusive mediation (Olawuyi, 2017). Elders' authority comes from representing communal norms and interests. This worldview emphasises consultation, consensus, and cooperation in governance over factionalism. Leadership means ethically mobilising communal resources and capabilities, not domination. African communitarian ethics provides vital resources for conceiving identity, law and politics in interconnectedness and shared duty. While openings exist to blend communal values with individual rights and pluralism, African thought powerfully articulates notions of personhood, morality, and order centred on human interdependence and dignity through community. However, prevailing governance paradigms struggle to integrate these foundations. Liberal democratic models enshrine atomised individual liberties over social obligations. Laws aim to deter deviant individuals, not heal collective relationships. Neoliberal policies impose market competition that severs communal bonds. Technocratic governance discounts consensus-building and care ethics. Nevertheless, African philosophies questioning these assumptions may suggest alternatives.

Personhood cultivated through reciprocal contribution to the community provides a counterpoint to self-interest. Shared access and collective advancement signify justice rather than just individual rights. Leadership as an ethical stewardship office contrasts with authoritarianism or totalitarianism. Restorative reconciliation healing rather than punitive justice aligns with maintaining social harmony. Consensus-oriented governance guided by care ethics diverges from adversarial political competition between factions (see Matolino, 2018). Ubuntu thinking envisions persons empowered by enacting communal roles, not pursuing narrow individualism. The African worldview articulates notions of identity, law, morality, and politics centred on human interconnection, ethics, and dignity through community participation. While openings exist to blend African principles with individual talents and pluralism, rediscovering indigenous communitarian foundations marginalised under colonialism remains vital amid contemporary atomisation. Their articulation of enduring collective selfhood and morality through social contribution provides indispensable resources to conceive cooperative technological futures benefiting all humanity. Even if synthesised with Western ideas, African communalist philosophy powerfully counters the anti-humanistic outcomes of market fundamentalism and technocratic control by re-centring communal values, reciprocity, and dignity.

African worldviews conceive human life as embedded in and dependent on the natural order (Mbiti, 1969). Sustenance comes from respectful co-existence with the environment. Honouring ancestral spirits believed to inhabit places like rivers, forests, and mountains maintains harmony between human needs and ecological limits (Tosam, 2019). This sacralisation of nature contrasts Western notions of the environment as resources for human exploitation. As Chemhuru (2019) notes, African knowledge bounded resource use within rituals and taboos to prevent over-extraction and restore equilibrium. Humanity belongs to nature, not apart from it. Agricultural and hunting practices aligning with seasonal rhythms reflect this interdependence ethics (Tosam, 2019). Thus, Humans did not create nature, so it deserves reverent care, not domination. People take only what is needful so nature's balance sustains generations. Proper relations between humans and the environment promote human flourishing. African environmental ethics provides grounding to develop technology aligned with, not exploitative of, natural systems.

Additionally, African humanism centres on ethics of care for human dignity beyond material utility (Metz & Gaie, 2010). Ubuntu embodies valuing all people unconditionally through empathy and reciprocity, with fulfilment coming from community contribution. Respect and compassion become primary motivations, not laws or abstract rules. This care ethic contrasts with Western legalistic individual rights and contracts. Care in African thought also extends to maintaining social harmony by resolving disputes (Adeyinka & Ndwapi, 2002). Peaceful conflict resolution matters more than punishment. Upholding communal unity takes priority over rigid rules when relationships can be repaired. However, African care ethics emphasises nurturing each member's welfare through inclusive deliberation on moral matters, not imposed norms. The goal is a cooperative moral community through synthesis, not coercive unanimity (Gyekye, 2004).

In contrast to Western individualism, African philosophy questions atomism's corrosive impacts on the community (Eze, 2008). Unchecked self-interest enables exploitation, conflict and frayed social ties. For Menkiti (1984), the detached individual defies human relational identity. African thought articulates duties to the collective over just individual rights (Gyekye, 1992). Responsibilities are accorded based on one's capability to contribute to society's welfare. This grounds social policies curbing inequality, unlike liberalism's tolerance of disparities (Howard, 1992). However, African thought also recognises individuality, critiquing conformist communalism suppressing dissent and diversity (Matolino, 2009). Personhood is shaped through community, attributes, and choices (Kaphagawani, 2000). Moderate communitarianism balances individual and collective needs by premising freedoms on not harming common interests (Gyekye, 1997).

5.4   An African Philosophy of Technology

African worldviews grounded in communal values, sacred ecology, and human dignity offer alternatives for conceptualising technology. From an African perspective, the tendency to reduce technology to efficiency and dominion risks severely severing human-environmental interdependence. An Afrocentric philosophy seeks technologies that foster balance within ecosystems as humanity's shared habitat. African knowledge systems have long integrated resource use within rituals and taboos that honour nature's limits, contrasting Western logic of unconstrained exploitation. Technology guided by African cosmologies would sustain nature's equilibrium rather than disrupt it. For instance, hydropower generation integrating photocells or turbines into flowing ecosystems respects planetary cycles. Biomimetic designs learn sustainability from natural forms and processes. Production can be re-embedded within circular ecological flows, minimising waste. African spirituality teaches that humanity belongs to the web of life—wisdom vital for guiding technology to harmonise with nature (Tosam, 2019).

African humanism centres communal values, designing technology first to serve society's shared needs and fulfil human dignity. This community orientation produces innovation-seeking collective advancement, counter to inequitable technoscience. Local priorities and participatory processes shape human-centred Afrocentric design. Community-led innovation makes technology more inclusive and appropriate than importing one-size-fits-all solutions (UN, 2018). For example, decentralised renewable energy systems based on village self-help traditions can expand electrification faster than distant state grids. Platform cooperatives and grassroots maker spaces foster co-creation, aligning technology with local contexts. Reciprocity, shared benefits, and care ethics guide technology development to strengthen social ties and equality. Tools expressing the solidarity and creativity of African communalism carry revolutionary potential to democratise technological futures. However, prevailing paradigms struggle to integrate these foundations. Hyper-individualism drives competitiveness, not cooperation. Atomised cost-benefit analysis discounts shared benefits. Products of exploitative supply chains damage both human and ecological communities. Yet African philosophies suggest more holistic, ethical alternatives aligned with communal well-being and environmental sustainability. African cosmologies articulating interconnection, care, spirituality, and communal values can reshape dominant paradigms of technology as mere tools of efficiency and control. They envision ecological harmony and collective well-being as technology's rightful goals. Even where synthesised with global knowledge, African humanistic principles are indispensable for guiding innovation to empower communities materially and spiritually.

Mainstream paradigms prioritise technology transfer and mimicry of foreign models as routes to development (Abdulai et al., 2022). However, standardisation often displaces sustainable indigenous knowledge and communal structures. Afrocentric philosophy questions the uncritical imposition of external technical universalism. Technologies localised through participatory adaptation to meet situated needs better serve African advancement. This positions users as experts shaping technology for their realities, not just passive consumers. For instance, we could create spaces fostering youth digital fabrication with local materials to enable context-specific innovation. We could also provide offline networks blending hyperlocal content with global knowledge to promote learning without dependence on standardised platforms. Such localised designs would allow diverse African identities and practices to thrive while selectively integrating external elements, moving from standardisation to sustainable self-determination. Afrocentric philosophy thus celebrates localisation as resistance to imperial diffusion.

Western theories portray technology as supplements to human deficiencies and limitations. However, Afrocentric thought envisions technology fulfilling humanity and dignity. This paradigm empowers users as creative agents, not just passive consumers. Tools designed for autonomy, understanding, and communal benefit empower African people materially and spiritually. Agricultural technologies enabling sustainable village self-reliance exemplify emancipatory models. Through cooperatives, digitally literate local youth can democratically govern platforms spreading capabilities. However, external technologies often disempower through opaque, proprietary designs inhibiting participation (Abdulai et al., 2022). Making innovation participatory requires platforms to enable users to become co-designers. Shared technological capabilities cultivated within communal ethics have deeply liberatory possibilities across Africa.

Under Western economic calculations, technological advancement risks deepening dehumanisation, inequality, and ecological ruin. However, re-centring humanistic African values can reorient technology from domination towards fostering care, dignity and justice. Ubuntu philosophy places proper relations between persons as technology's rightful purpose. Tools expressing African communalism's solidarity and creativity have emancipatory potential. Digital systems could build trust and reciprocity rather than division. Replacing efficiency metrics with social welfare goals reorients innovation to heal rather than disrupt communities. A moderate communitarian approach would balance individual and collective needs under moral guidance. Shared abundance guided by African socialism would enable technology for creative works benefiting all. Beyond material goals, technologies can fulfil spiritual meaning-making, while interdependence with nature sustains prosperity. Re-grounding technology design in African relational ethics sustains both innovation and dignity.

Mainstream Western paradigms prioritise productivity, efficiency, precision, and the rationally exercised domination of nature and society as technology's goals. Tools are viewed as value-neutral and autonomous means for individuals to maximise personal benefit and gain. Mastery over an externally conceived nature to minimise dependence and extract resources is considered inherent progress (see Fig. 5.1). The competitive advancement of individuals and firms trumps collective needs. Technocratic metrics of functionality and economics guide the assessment of technology rather than social and ecological impacts. In contrast, the African conceptual principles generally emphasise collective human well-being, harmony with nature, and the intrinsic value of human dignity. Technology is framed as an instrument for fulfilling communal needs and enhancing social bonds beyond maximising material productivity and efficiency. Innovation is a means for sustaining interdependence with the natural world through circular, regenerative systems rather than exerting mastery over nature through disruption and exploitation (see Fig. 5.1).

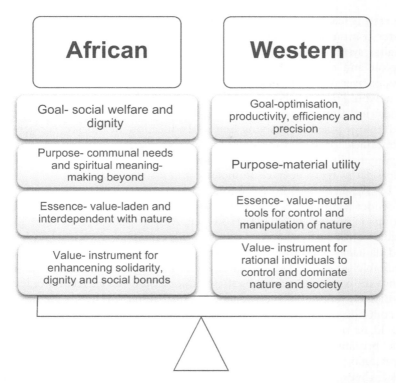

Fig. 5.1 Contrast between African and Western conceptions of technology

Democratised, cooperative advancement represents progress, not technocratic control and optimisation.

5.5 Applying African Thought to Technology

Pragmatic principles rooted in Africa's wealthy cultural foundations may offer vital guidance on shaping tools for holistic advancement. Under capitalist paradigms, technological progress is measured by productivity, economic growth and material accumulation rather than improvements in holistic human development enabling dignity and flourishing. However, African thought defines the purpose of technology as advancing human welfare within egalitarian, ecological, and spiritual dimensions. Adopting the Ubuntu philosophy would orient innovators to design tools expressing care, reciprocity, and interdependence. Technologies would foster community, solidarity, and reconciliation over control or violence. Digital systems could build trust and decolonise rather than distract, exploit and divide. Medical advances would seek health equity for all groups over profit maximisation. Infrastructures like roads and electrification would aim to benefit collective prosperity rather than extractive industries. Equally, policymakers applying African socialism would assess technology based on enabling universal education, healthcare, cooperative economics, participatory democracy, and societal betterment beyond aggregates like GDP. Technology designed first to uplift human dignity offers pathways to balance abundance and justice.

In Western economics, unregulated markets and utilitarian ethics engender technologies that exacerbate inequalities and harm marginalised groups (Popa et al., 2023). However, African philosophies offer moral guidance for justly sharing benefits while minimising technology's risks and damages. Communal values of interdependence and care emphasise equitable access and democratising technology as collective resources. Innovation systems guided by these ethics would ensure technologies like medicines, renewable energy, and agricultural improvements are oriented to common welfare rather than privatisation. Intellectual property regimes would be reformed to enable knowledge sharing to meet urgent needs. The precautionary principle would also guide restricting technologies threatening well-being based on communal discernment of acceptable risks. Emerging technologies like geoengineering, genetically modified organisms, and artificial intelligence would face stringent assessments of social impacts and moral implications guided by African wisdom traditions. This foresight would direct adaptation to prevent disruptions to the social fabric.

Western legalistic approaches tend to delimit technology ethics within rigid rules, patents, and proprietary constraints. However, African humanism provides a foundation for participatory, consensus-based design processes actualising cooperative moral agency. Ubuntu thinking recognises technological artefacts as expressions of shared values. Instruments like medicines, digital platforms, and infrastructure designed through collective consultation and transparency better align with local contexts and needs. Engineers and policymakers collaborating directly with communities gain insights to shape appropriate solutions. Iterating prototypes enables context-specific learning absent in standardised models. Further, participatory design allows technology users to become co-creators by bringing local knowledge, priorities, and creativity into democratic innovation. Technoscience guided by reciprocal ethics ceases to be alienating or exploitative. The aim becomes innovation reflecting African identities rather than universal imposition. Such communal pragmatism reconciles modern tools with cultural contexts.

Additionally, African thought provides value-centric foundations to assess technology based on its contributions to communal well-being and sustainability. The capability approach centred on African socialism would judge innovations by their role in enabling universal education, healthcare, cooperative economics, and participatory democracy—substantive human development metrics rather than just aggregates like GDP. Equally, African environmental ethics provide frameworks to evaluate technology based on preserving sacred ecologies and sustaining future generations. Assessing tools by their alignment with circular material flows and bolstering community self-reliance elevates sustainability. Shifting focus from technology's quantity, reach, or monetisation towards quality of life and social impacts fosters humanistic innovation. Policy reforms guided by African thought's insights on the right relationships and holistic progress are vital for just, inclusive technological futures.

The African principles generally emphasise localised, grassroots, cooperative and participatory technology applications. They envision technologies adapted through inclusive, bottom-up processes to fit specific cultural contexts and community needs. Innovation is framed as a collective endeavour rather than an imposition by distant experts and institutions (see Fig. 5.2). To serve shared interests rather than capital accumulation, technologies are designed to be governed cooperatively by ordinary users and citizens. Access is democratised through participatory methods, empowering marginalised groups traditionally excluded from shaping technology development and policy. Further, integrating indigenous and modern technological systems through participatory processes can enable complementary advancement of inclusive social and developmental goals. Thus, blending external innovations with contextual adaptation sustains communal knowledge while appropriating science to solve localised needs.

5.6   Envisioning New Technological Paradigms

Modernity's narrow aim for technology has been ever-greater efficiency, precision, and control of nature and society. However, optimising technical means often lose sight of ethical ends. Unquestioned 'progress' risks

Fig. 5.2 African philosophical principles in design and use of technology

engendering dehumanisation, inequality, and environmental ruin. Alternative paradigms centred on African humanism consider technology's social and ecological impacts above productivity or profitability. The innovation aim is to fulfil human dignity within communal and environmental balance. Engineering guided by care ethics and sustainability constrains tendencies towards domination through technology. Progress is redefined as equitable prosperity in harmony with living systems. Efficiency metrics shift from monetary valuations and output towards social welfare and ecological regeneration. Participatory governed technology answers marginalised communities' needs on their own terms. The mindset moves from seeking control towards creative flourishing.

Western market logic engenders excessive patent protections and proprietary constraints incongruent with technology's collective knowledge foundations. However, African communalism provides grounding for cooperative innovation and economics, seeking open access and shared benefits. Governance innovations like stakeholder trusts, open-source licensing, and collaborative design can direct technology to serve inclusive prosperity. Platform cooperatives enable grassroots digital services owned and controlled by local users, not distant corporations. Socialised medicine and participatory assessment guided by African ethics can democratise decisions. Local manufacturing, farming, and energy cooperatives pool community resources to meet needs sustainably without dependence on exploitative global markets. The aim becomes mutual flourishing through solidarity and reciprocity, not individualistic profit maximisation. African socialism provides models to advance science as creative works for collective abundance.

Due to the inequitable diffusion and design of modern technologies, social participation and civic engagement capacities remain stratified by gender, class, geography, and other axes of marginalisation (Hyytinen & Toivanen, 2011). However, policies guided by African communal principles can democratise technological access, literacy, and participation. Structural reforms are needed to decolonise innovation ecosystems that stifle African knowledge, initiative, and self-reliance. Grassroots skill development and decentralised manufacturing enable wider appropriation of technologies like renewable energy, sustainable farming, and digitally enabled services on communal terms. Public maker spaces and digitisation of indigenous knowledge systems foster bottom-up innovation. However, democratisation requires going beyond technical access to enable co-creation. Platforms channelling user creativity invert consumption towards participatory design. Education that integrates technical and ethical learning cultivates engaged citizenship (Grad & van der Zande, 2022). Enriching technological literacies through African wisdom traditions liberates new social possibilities.

Envisioning more just technological futures requires embracing epistemic diversity and inputs from historically marginalised knowledge systems. Calls for 'ethical technology' will remain shallow, absent contributions of African philosophy. However, centring on African humanistic values of care, dignity, and the collective good can help morally orient technology globally. The Ubuntu paradigm would infuse technology design and policy worldwide with relational ethics seeking cooperative advancement rather than domination. Sacred ecologies guide sustainable innovation within planetary boundaries. Economic paradigms shift from individualism towards collectivism. Shared principles of social justice and care ethics drawn from the wellspring of African thought can direct technology as a force for universal human flourishing, not destruction. Freeing technological imagination from the narrow strictures of Western philosophy opens possibilities for pluralistic, ethical innovation across diverse contexts. Africa's conceptual resources contain wisdom vital for our shared planetary future. However, paradigm change requires elevating subjugated voices to reshape the foundations of how we design and govern technology. African thought provides indispensable guidance for just and ecological technological futures by guiding innovation to care for nature and society.

Realising these possibilities demands urgent action. Policies can foster communal technology models like cooperatives and participatory design to invert consumption and empower creative users. Public investments can expand grassroots innovation ecosystems and local manufacturing capabilities across Africa, ending reliance on foreign expertise and charity. Reformed curricula integrating African knowledge traditions build capacities to leverage technology for cooperative advancement. Regional networks facilitate knowledge exchange on leveraging African innovation philosophies. Consensus governance institutions uphold collective oversight and adaptation of technologies to prevent disruption. International discourses on the ethics of emerging technologies should centre African perspectives on humanism, dignity, and ecology. Accessibility regulation mandates inclusive design principles. Rights of nature provisions constrain unaccountable technical exploitation. Genuine technological self-determination requires decolonising the philosophical foundations of innovation. Recognising Africa's marginalised knowledge heritage reframes technology as an instrument of care and justice.

5.7   Conclusion

This chapter has examined dominant Western and African conceptual frameworks of technology, revealing needs and possibilities for transforming prevailing innovation paradigms. Mainstream assumptions valorise technology as an apolitical, autonomous process of value-neutral tools for control and mastery over nature and society, ignoring ethical values, social impacts, and cultural contexts. However, African thought traditions articulating humanistic principles of communalism, sacred ecology, care ethics, and human development offer alternative foundations for conceptualising, evaluating, and governing technology in more empowering, ethical and socially just ways. The instrumental Western paradigms problematically reduce nature and society to raw materials for exploitation, peoples to nodes in efficiency systems, and technological advancement to scientifically exercising domination and precision control absent moral guidance. These ideologies risk fostering reckless instrumentalisation of the planet and people, undermining collective fabrics sustaining human dignity. However, African philosophies provide vital conceptual resources and pragmatic guidance for re-centring ethical responsibility, ecological sustainability, democratisation, and care for human capabilities in technological thinking and policymaking.

This interrogation of background assumptions and oppressive potentialities of technology detached from wisdom aims to spur discussion on transforming education, policy, assessment, and design processes. Genuinely empowering and socially beneficial technological futures require centring the knowledge and perspectives of marginalised peoples excluded from designing mainstream systems that optimise control for elite interests. African orientations towards community, care, and humanistic principles highlight alternative pathways to shape innovation for collective advancement and purposeful harmony with nature. Cognitive shifts are still required to overcome the dominant paradigms' narrow technical rationality and blind spots to systemic inequalities. However, African thought contributions can help equip reform efforts with different premises and priorities. Beyond just better technical functionality and efficiency, technology aligned with human dignity, capabilities, and ecology is possible. This chapter has demonstrated that extracting conceptual resources and pragmatic principles from African philosophy would empower efforts to realise more ethical, empowering technological futures benefiting humanity.

References

Abdulai, A., Murphy, L., Thomas, A., & Thomas, B. (2022). Technology transfer offices and their role with information mechanisms for innovation performance in firms: The case of Ghana. *Knowledge, 2*(4), 719–734. https://doi.org/10.3390/knowledge2040041

Adeyinka, A. A., & Ndwapi, G. (2002). Education and morality in Africa. *Pastoral Care in Education: An International Journal of Personal, Social and Emotional Development, 20*, 17–23. https://doi.org/10.1111/1468-0122.00225

Bacon, F. (1920) [1964]. *On the interpretation of nature or a science of productive works.* (B. Farrington Trans.). Liverpool University Press.

Brey, P. (2010). Philosophy of technology after the empirical turn. *Techné: Research in Philosophy and Technology, 14*(1), 36–48.

Chemhuru, M. (2019). The moral status of nature: An African understanding. In *African environmental ethics: A critical reader* (pp. 29–46). Springer Verlag.

De Vries, M. J. (1997). Science, technology and society: A methodological perspective. In M. J. De Vries & A. Tamir (Eds.), *Shaping concepts of technology*. Springer. https://doi.org/10.1007/978-94-011-5598-4_3

Drengson, A. R. (1982). Four philosophies of technology. In C. Hanks (Ed.), *Philosophy today* (pp. 26–37). Wiley-Blackwell.

Eze, M. O. (2008). What is African communitarianism? Against consensus as a regulative ideal. *South African Journal of Philosophy, 27*(4), 386–399.

Fallman, D. (2011). The new good: Exploring the potential of philosophy of technology to contribute to human– Computer interaction. In *Proceedings of the SIGCHI Conference on Human Factors in Computing Systems (CHI' 11)* (pp. 1051–1060). Association for Computing Machinery. https://doi.org/10.1145/1978942.1979099

Famakinwa, J. O. (2010). The moderate communitarian individual and the primacy of duties. *Theoria, 76*(2), 152–166.

Feenberg, A. (1991). *Critical theory of technology*. Oxford University Press.

Feenberg, A. (1995). *Alternative modernity: The technical turn in philosophy and social theory*. University of California Press.

Feenberg, A. (2002). *Transforming technology: A critical theory revisited*. Oxford University Press.

Ferré, F. (1988). *Philosophy of technology*. University of Georgia Press.

Gardner, P. L. (1997). The roots of technology and science: A philosophical and historical view. In M. J. De Vries & A. Tamir (Eds.), *Shaping concepts of technology*. Springer. https://doi.org/10.1007/978-94-011-5598-4_2

Grad, J. M., & van der Zande, I. S. E. (2022). Cultivating global citizenship through higher education: A reflection on the development from civic to global engagement. *Education Sciences, 12*(11), 766. MDPI AG. Retrieved from. https://doi.org/10.3390/educsci12110766

Gunnell, J. G. (1982). The technocratic image and the theory of technocracy. *Technology and Culture, 23*(3), 392–416. https://doi.org/10.2307/3104485

Gyekye, K. (1992). Person and community in African thought. In K. Wiredu & K. Gyekye (Eds.), *Person and community: Ghanaian philosophical studies* (Vol. 1, pp. 101–122). Council for Research in Values and Philosophy.

Gyekye, K. (1997). *Tradition and modernity: Philosophical reflections on the African experience*. Oxford University Press.

Gyekye, K. (2004). *Beyond cultures: Perceiving a common humanity* (p. 195). The Council for Research in Values and Philosophy.

Hanks, C. (Ed.). (2010). *Technology and values: Essential readings*. Wiley-Blackwell.

Howard, R. E. (1992). Communitarianism and liberalism in the debates on human rights in Africa. *Journal of Contemporary African Studies, 11*(1), 1–21.

Hyytinen, A., & Toivanen, O. (2011). Income inequality and technology diffusion: Evidence from developing countries. *The Scandinavian Journal of Economics, 113*(2), 364–387. http://www.jstor.org/stable/23016840

Kaphagawani, D. (2000). African conceptions of personhood and intellectual identities. In P. H. Roux (Ed.), *The African philosophy reader* (pp. 240–244). Routledge.

Lemmens, P., Blok, V., & Zwier, J. (2017). Toward a terrestrial turn in philosophy of technology Guest editor's introduction. *Techne: Research in Philosophy and Technology, 21*(2–3), 114–126.

Lettow, S. (2011). Somatechnologies: Rethinking the body in philosophy of technology. *Techne: Research in Philosophy and Technology, 15*(2), 110–117.

Marcuse, H. (1964). *One-dimensional man: Studies in the ideology of advanced industrial society*. Routledge.

Matolino, B. (2009). Radicals versus moderates: A critique of Gyekye's moderate communitarianism. *South African Journal of Philosophy, 28*(2), 160–170.

Matolino, B. (2018). *Consensus as democracy in Africa*. NISC.

Mbiti, J. (1969). *African religions and philosophy*. Doubleday and Company.

Menkiti, I. A. (1984). Person and community in African traditional thought. In R. A. Wright (Ed.), *African philosophy: An introduction* (pp. 171–181). University Press of America.

Metz, T., & Gaie, J. B. R. (2010). The African ethic of Ubuntu/Botho: Implications for research on morality. *Journal of Moral Education, 39*(3), 273–290. https://doi.org/10.1080/03057240.2010.497609

Mitcham, C. (1985). What is the philosophy of technology? *International Philosophical Quarterly, 25*(1), 73–88.

Olawuyi, D. S. (2017). Barriers to technology assimilation in Africa. In *From technology transfer to technology absorption: Addressing climate technology gaps in Africa* (pp. 3–5). Centre for International Governance Innovation. http://www.jstor.org/stable/resrep15517.7

Popa, E. O., Blok, V., Katsoukis, G., & Schubert, C. (2023). Moral impact of technologies from a pluralist perspective: Artificial photosynthesis as a case in point. *Technology in Society, 75*, 102357. https://doi.org/10.1016/j.techsoc.2023.102357

Tosam, M. J. (2019). African environmental ethics and sustainable development. *Open Journal of Philosophy, 9*, 172–192. https://doi.org/10.4236/ojpp.2019.92012

UN. (2018). *Mainstreaming inclusive technology and innovation policies that leave no one behind*. Economic and Social Commission for Asia and the Pacific.

van Niekerk, J. (2007). In defence of an autocentric account of Ubuntu. *South African Journal of Philosophy, 26*(4), 364–368. https://doi.org/10.4314/sajpem.v26i4.31494

Winner, L. (1997). Technology today: Utopia or Dystopia? *Social Research, 64*(3), 989–1017. http://www.jstor.org/stable/40971195

Wiredu, K. (1980). *Philosophy and an African culture.* Cambridge University Press.

CHAPTER 6

An African Social Constructivism of Technology

6.1 Introduction

This chapter looks at how social constructivist perspectives can be useful lenses for understanding the complicated ways that social systems, relationships, interests, meanings, and cultural values guide the development and use of technologies in Africa. Constructivism offers a different view than the common belief that technologies just follow their own technical path based only on scientific discoveries and engineering goals. It sees technologies as being shaped through many interactions between social and technical things, not as ready-made or neutral products. Social constructivism includes theories and ideas that technology is situational and cultural, not objective or universal as if it only follows scientific logic and efficiency (Pannabecker, 1991). Some key constructivist concepts are interpretive flexibility—how people understand technologies differently; technological frames——the beliefs and norms people have about technology; closure and stabilisation——how debates over technology get settled; large technological systems and networks; and the co-construction of technology and society. This view sees technologies arising from social processes, relationships, interests, and values as more than just technical factors.

Constructivist views diverge from traditional 'technical rationalist' perspectives on technology, which are still commonly held in many policy and development spheres. These problematic perspectives include assumptions

Y. Ndasauka, *African Mind, Culture, and Technology*,
https://doi.org/10.1007/978-3-031-62979-2_6

that technology follows internal scientific logic and technical optimisation imperatives rather than being shaped by cultural values or political agendas. Additionally, they presume artefacts and systems are value-neutral tools uniformly applicable across diverse contexts. There is also a technological determinism whereby technology is seen as unilaterally impacting cultures and societies based on its inherent technical qualities. This chapter contends that technologies in Africa should be understood as socially shaped artefacts and socio-technical systems that are embedded within and dynamically co-evolve with particular cultural contexts. Their design and integration trajectories inherently reflect the contested goals, assumptions, interests, and values of different empowered groups who are able to set strategic priorities within surrounding ecosystems of regulations, infrastructures, norms, and discourses. Indeed, technologies often extend the ideological visions and structural relationships of dominance found in the very social worlds in which they are funded, constructed, and deployed rather than following universal technical imperatives. However, they also demonstrate interpretive flexibility in how social groups may alternatively appropriate them towards varied agendas. Users retain some agency to deliberately reshape or localise technologies when their embedded biases, limitations, and disruptions manifest through lived experience. However, this frequently requires contesting constraints around access, literacy, participation, and ownership in order to realign innovations with alternative priorities. Ethical technology shaping demands elevating marginalised voices in order to transform tools designed primarily for elite interests into more empowering, humanistic instruments for collective advancement.

This chapter is structured as follows: Firstly, it demarcates constructivism from prevailing discourses of technological determinism. Secondly, it interrogates the interpretive flexibility of technologies within African contexts. Thirdly, it investigates multidirectional shaping processes between emergent technologies and evolving African societies. Fourthly, it explores the potential for participative co-construction of innovations aligned to localised requirements. Fifthly, it underlines the imperative to amplify plural African voices and priorities in these processes. The chapter concludes that Africa's technology should be situated and guided by these constructivist insights. The chapter suggests conceptualising African technology policies and innovation pathways to advance human development rooted in indigenous values rather than imported paradigms.

6.2   THE SOCIAL CONSTRUCTION OF TECHNOLOGY

Social constructivism views technologies as socially constructed through multidimensional interactions between diverse elements, including science, engineering, design, infrastructure, cultural meanings, economic forces, and political agendas. Technologies emerge from socio-technical environments rather than being predetermined by scientific advances or technical efficiency imperatives (Pannabecker, 1991). Critical concepts in constructivist theory include technological frame and interpretive flexibility, closure processes, and technological ensembles.

Technological frames refer to the goals, materials, practices, and cultural meanings that shape how technology is interpreted and designed by different relevant social groups (Bijker, 1995). Each technological frame comprises the intentions, values, norms, knowledge, and constraints through which a community approaches and gives meaning to a technology. Comparing frames reveals differing priorities and assumptions between inventors, users, regulators, etc. Mapping frames is crucial to see how technologies embed varied agendas. Technological frames reveal how goals, materials, practices, and cultural values shape technology development among different groups. On the other hand, interpretive flexibility means technologies hold diverse social meanings depending on cultural context and applications (Bijker et al., 1987). As innovations emerge, designs remain open to interpretation and negotiation by stakeholders. Different social groups appropriate technologies in multiple ways. A technology's meaning and conventions do not automatically flow from technical capacities but are constructed through social interaction and sensemaking.

Constructivism also recognises closure processes whereby flexibility reduces over time as dominant conventions and standards emerge. Closure refers to how the interpretive flexibility of technology reduces over time as shared conventions, dominant designs, technical standards, and cultural understandings stabilise across groups (Bijker & Law, 1992). It highlights the co-construction of technology and society through reciprocal shaping. It articulates how technologies become embedded within broader interdependent systems and infrastructures. Constructivism recognises technologies as products of social processes, relationships, interests and values rather than predetermined or value-neutral. This contrasts with technologically determinist views that technologies follow autonomous technical trajectories shaped by science alone and impact society unilaterally. The constructivist lens reveals technology as embedded in and co-shaped by cultural contexts, institutions, and agentic choices. This conceptual orientation holds essential implications for democratising innovation. Mapping closure dynamics reveals technologies stabilising through social processes, not just technical optimisation.

Technological ensembles refer to interdependent socio-technical systems where technologies become embedded within broader institutional structures, practices, infrastructures, supply networks, cultural norms, etc. (Hughes, 1987). Innovations shape and are shaped by linkages with transportation, energy systems, regulations, economic forces, consumption habits and political priorities. Ensemble theory reveals co-evolution between technical and social elements. Co-construction processes recognise innovation emerging from multidirectional interactions between new technologies and the social world as they reciprocally shape each other through use, design, regulation, meaning-making and integration (Fairchild & Quansah, 2007). Neither technical nor social systems unilaterally determine outcomes.

The intellectual foundations of social constructivist perspectives on technology evolved over decades through the seminal contributions of scholars across philosophy, history, sociology, and science and technology studies. Constructivist approaches arose as a compelling counterpoint to technologically determinist models, which viewed technologies as developing autonomously based on scientific advances and technical optimisation to impact society unilaterally (Fairchild & Quansah, 2007). Thinkers like John Dewey and Karl Marx laid the necessary philosophical groundwork for constructivist concepts during the late nineteenth and early twentieth centuries. Dewey's pragmatist philosophy emphasised technology's cultural situatedness and dynamic, problem-solving nature rather than viewing it as applied science (Hickman, 1992). Dewey also highlighted co-shaping relationships between tools and their users. Marx identified close linkages between developing technologies and modes of economic production within materialist class struggles, deviating from progressive assumptions (Mcquarie & Amburgey, 1978). These foundational perspectives recognised technology's social context beyond technical realms.

In the 1930s–1990s, scholars of the sociology of technology like Jacques Ellul, Lewis Mumford, and Langdon Winner further explored relationships between technology and society. Ellul (1964) examined how technological thinking shaped modern institutions and ideology. Mumford

(1934) provided a historical analysis revealing links between authoritarian social power and unchecked technics. Winner (1980) investigated how technologies can encode political biases and values in their designs, illustrating non-neutrality. These studies seeded constructivist approaches. The emergence of academic fields like science and technology studies (STS) in the 1980s provided rich empirical research demonstrating technology's contingency on social processes, cultural contexts, and power dynamics rather than deterministic technical imperatives. For instance, Wiebe Bijker's research on bicycle designs illustrated the critical concept of interpretive flexibility—the diverse social meanings, goals and negotiated conventions shaping technologies (Pinch & Bijker, 1984). Feminist STS scholars revealed embedded gender biases in areas from transport systems to medical devices. Constructivist analysis illustrated how technologies extend dominant social groups' ideologies and structural interests.

A related intellectual stream in STS was an actor-network theory developed by scholars like Bruno Latour, John Law, and Michel Callon, which mapped innovations arising from networked negotiations between diverse human and nonhuman nodes. This approach dissolved false dichotomies between technology and society by highlighting hybrid socio-technical actants and relationships. Constructivist perspectives arose over decades through the critical, empirical insights of philosophers, historians, sociologists and STS scholars converging in recognising technology as socially shaped and culturally embedded innovations. This knowledge foundation continues informing current technology analysis and governance frameworks.

Social constructivist perspectives differ fundamentally from traditional technical rationalist understandings of technology in several key dimensions. Constructivism emphasises technology's contingent and culturally embedded nature rather than viewing it as an objective, universal or value-neutral domain driven by scientific logic and technical optimisation (Pannabecker, 1991). Technological rationalism views technology development as an autonomous process guided by internal scientific advances, efficiency imperatives and engineering problem-solving. New technologies are assumed to emerge from expert communities in a linear pattern building upon prior technical knowledge (Feenberg, 1991). In contrast, constructivism reveals technology development as a 'heterogeneous engineering' process richly shaped by its socio-cultural context, including economic and political agendas, institutional norms, public discourses, designer assumptions, user interactions and infrastructure linkages (Law, 1987). Innovation unfolds through multidirectional co-construction between the technical and social worlds.

Technical rationalism presumes technologies are value-neutral tools applicable without bias across contexts. Technological thinking is an objective, apolitical reasoning about instrumentally optimal means. Constructivism counters this, demonstrating how technologies embody and propagate their designers' and sponsors' values, interests, ideologies, and assumptions (Winner, 1980). Innovations are political artefacts shaped within cultural beliefs, hierarchies and conflicts rather than impartial tools. Feminist and postcolonial STS scholarship reveals deep exclusions and power biases encoded into dominant technologies and scientific paradigms. Technological determinism views technology as an autonomous force impacting cultures and societies unilaterally based on its inherent technical qualities. Social transformations are seen as predominated by technology. Constructivism reveals technology's social impacts and meanings as contingently produced through reciprocal interactions between technologies and heterogeneous social systems (MacKenzie & Wajcman, 1985). Social forces actively shape how technology is interpreted, regulated, and integrated in context. Recognition of co-production opens space for democratising innovation pathways.

Technical rationalism presumes technological solutions developed in Western industrial contexts as universally optimal and transferable across global settings with uniform effects. Constructivism recognises diverse technological frames, use patterns, and impact variances across cultural contexts. It emphasises situated specificity over universalism in socio-technical change. Constructivism's richer recognition of contingency, biases, diversity, and co-creation provides crucial guidance for more democratically governing technology futures in contrast to decontextualised assumptions of technical rationalism. This conceptual shift remains imperative.

6.3   Technology Shaped by Social Systems

The trajectory of technology design and integration in society follows directional pathways shaped by the particular agendas, values, interests, and ideologies of influential groups empowered to set strategic priorities and mobilise resources within institutional ecosystems. Innovation thus aligns with the goals and social visions of those with cultural capital and structural power rather than as an undirected apolitical technical process. A salient example is the shaping of post–World War 2 research and innovation systems in the United States under the dominant agenda successfully promoted by Vannevar Bush, director of the US Office of Scientific Research and Development. Bush's 1945 report titled 'Science: The Endless Frontier' laid out an influential vision oriented towards advancing national security, economic growth, global power projection, and ideals of scientific control over nature by structuring innovation ecosystems around a close university-government-military alliance underpinned by generous public research funding. This particular set of priorities drove the trajectories of American industrial and military technologies for decades as Bush mobilised resources and institutional reforms to align research agendas with his articulated vision.

Similarly, modern consumerist philosophies valuing disposability, individualism, and materialism as paths to happiness have shaped the development of convenient but wasteful technologies like fast fashion garments and single-use plastics by aligning innovations with particular cultural goals (Bick et al., 2018). In contrast, visions of human harmony with nature have driven movements like appropriate technology and permaculture design seeking more ecologically regenerative and socially empowering technical solutions aligned with alternative progressive values (Willoughby, 1990).

The priorities and assumptions encoded into institutions like corporations, universities and governments through policies, funding budgets, grants and infrastructure shape which technologies are advanced. For instance, the Green Revolution's fertiliser-intensive agricultural paradigm focusing on cereals and responsive varieties displaced alternative approaches aligned with smallholder ecological knowledge and diverse crops by institutionalising a particular modernisation vision (Glover et al., 2016). Technology choices often emerge from contests between competing values and interests seeking institutional encoding. Explaining the full range of motivations, biases and exclusions underlying particular agendas is vital.

Recognising technology as inherently shaped by special interests provides an impetus for democratising innovation by opening priority-setting and resource mobilisation to more plural, inclusive, and ethical visions of progress, human development, sustainability, and well-being. Africa's technological future, in particular, lies in elevating diverse endogenous perspectives to steer change rather than relying on narrow technical paradigms exported from the Global North.

Applying a moderate communitarian lens reflecting African communal principles reveals the dynamics of technologies transforming and being transformed by social structures. Unlike radical communitarian negation of individual interests, moderate communitarianism acknowledges legitimate personal aspirations within ethical limits, valuing collective well-being (Gyekye, 1997) This aligns with African humanism, balancing individual and community interests and goals. From this perspective, technologies fostering solidarity, empathy, and cooperation enable human development, while innovations advancing selfishness and social fragmentation violate ethical social bonds. Though technologies interact with and influence all spheres of life, consciously shaping them for communal good is possible through mindful guidance and governance. For instance, innovations like mobile money and village solar microgrids have strengthened communal finance and energy access, reflecting cooperative values. However, uncontrolled digital media risks fuelling dangerous individualism if not appropriately directed. Communities proactively localising technologies like farming techniques or digital storytelling can align tools with shared visions. The challenge for African societies is consciously steering rising technophilic tendencies towards humanistic paths that avoid extremes of destructive communalism or atomistic materialism. With mindful, ethical shaping, technologies can be levers for finding a balance between legitimate personal and collective aims.

Looking deeper, subtle communitarian-technicist tensions become visible. As Murove (2009) explained, African communitarianism integrates social and ethical dimensions largely lacking in Western individualism. Relationships, duties, and human dignity take priority over profits and material gains. However, the implicit logic of technological systems often promotes impersonal materialist values antithetical to humanistic communitarian worldviews. Uncritical adoption of such technologies can corrode communal bonds. For instance, while agricultural machinery can ease farming burdens, thoughtless mechanisation displaces village labour and ruptures rural social fabrics.

Similarly, digital platforms like Facebook may connect friends through convenience but also breed isolated virtual bubbles that neglect offline communal ties (Ndasauka, 2021). Entrepreneurial success stories celebrated in social media often inflame greed and unilateral wealth accumulation detached from the community. Therefore, requisite wisdom entails discerning technologies aligned with ethical, humane ideals from those advancing selfish interests contrary to the common good. This demands a nuanced illustration of tensions between techno-capitalist drives and communalist philosophies. As Gyekye (1997) illustrated, African communitarianism recognises certain individual rights correlative to social duties. Here, technologies respecting legitimate personal and collective interests can enable human development. For example, when appropriately guided, mobile telephony allows convenient communication aligned with cooperative relationships and mutual obligations. However, the same technologies devoid of ethical mooring can become tools of deception, fraud, and dangerous disconnection.

Moderate communitarianism eschews extremes, seeking principled balance. The task for African societies is steering rising technologies through this middle path between radical collectivism abolishing individuality and radical individualism neglecting communal imperatives. With conscientious oversight and regulation, emerging innovations like digital finance platforms can strengthen financial inclusion and communal economic empowerment. However, the same digital tools devoid of oversight risk predatory lending violating consumer protections or cryptocurrency speculation eroding financial stability. Therefore, mindful localisation and governance of technologies remain essential. This nuanced dance of calibrating technological tools and systems with moderate communitarian philosophies will likely grow in importance as emerging technologies increase across Africa. Advanced innovations like artificial intelligence and gene editing already compel ethical reckoning, as their misapplication may go against some African values. Therefore, thoughtful adaptation within communal-ethical bounds remains vital. Ultimately, the hope is for African societies to harness technologies for human development goals by balancing material advances with spiritual, moral, and communal priorities. This demands ongoing discernment and deliberation to harmonise innovations with African humanistic values. The promise is great if technology is guided as instruments for mutual uplift rather than uncontrolled currents eroding social cohesion.

Technologies interact dynamically with African communities' diverse social roles, groups and demographics, revealing equity questions surrounding differential access and adoption. New technologies like computers, smartphones, and software initially spread disproportionately among educated, urban middle-class professionals rather than marginalised rural demographics lacking resources and tech literacy. However, broader general-purpose applications like basic mobile phones help bridge access gaps (Porter et al., 2015). Youth frequently embrace new technologies like social media, digital finance platforms, or crypto assets more eagerly than elders, spurred by peer adoption, aspirational cultural messaging and greater comfort in adopting innovations. However, wisdom from lived experience remains essential for communities in discerning the appropriate integration of new tools. While youth may adeptly manoeuvre emerging gadgets and apps, elder guidance is crucial to anchor technological exuberance in ethical wisdom traditions.

Gender gaps persist but can gradually close if conscious efforts are made to provide women and girls equal opportunities and technology education, as demonstrated by women effectively leveraging technologies like e-commerce, mobile money and digital networking for economic empowerment once structural and cultural constraints are dismantled (Pesando & Rotondi, 2020). However, well-designed interventions can unlock technology's potential to empower women as entrepreneurs, connect them to health services, and amplify their voices in civic debates. Ethnic, religious, and linguistic minorities face discrimination barriers that can exclude them from technological opportunities without thoughtful correctives. Comprehensive localisation of technology infrastructure, content, and interfaces can help incorporate marginalised groups by enhancing tool relevance.

People with disabilities represent another constituency requiring inclusive technology design and accessibility provisions to ensure technologies empower rather than inhibit their aspirations (Moon et al., 2019). Assistive technologies tailored for people with disabilities, like screen readers, voice interfaces, and tactile apps, when designed sensitively, grant greater independence and dignity. Indigenous communities need technologies adapted to their unique contexts, capacities and cosmovision rather than disruptive tools violating their cultures. Approaches respecting indigenous self-determination over adoption choices allow for harnessing innovations on local terms. Careful and participatory shaping of technological innovation is required to appropriately empower disadvantaged segments of society rather than exacerbating exclusion and inequality. Avoiding a narrow focus only on elite capture necessitates consulting diverse community voices and priorities in co-constructing technological futures. Such localisation enables culturally resonant adaptation and participatory priority-setting, aligning technologies with communal worldviews for collective advancement. With thoughtful governance and social conscience, Africa's unfolding technological transformations can equitably empower the continent's diverse populace.

As new technologies diffuse in African societies, they interact dynamically with established social institutions, spurring adoption and adaptation processes that reveal local priorities negotiating external paradigms. In government administration, innovations like enterprise software, biometric identification systems and data analytics have rapidly transformed bureaucracies with digitisation, promising efficiency gains in service delivery but posing risks of automating corruption or deteriorating accountability. Thoughtful oversight is essential to ensure administrative technologies enhance good governance and citizen empowerment rather than worsening public sector opacity and abuse. Emerging financial technologies like mobile money and blockchain are rewiring African banking ecosystems to become phone-based rather than branch-based. This technology-enabled credit scoring expands access but requires consumer protections against predatory lending.

Healthcare is transforming through telemedicine, which provides remote consultations and assistive apps to manage time and patient data. Digital health solutions enhance access but must include marginalised communities through intentional design. Education rapidly integrates online learning platforms, presenting opportunities to expand access amid the pandemic but requiring diligent equity safeguards against a digital divide. Dynamic technology debates continue around potentially disruptive innovations like electoral biometrics, genetically modified crops, platform economies, workforce automation and artificial intelligence, which compel deliberation on social benefits and risks before adoption. Such complex and politically charged innovations require inclusive democratic processes, allowing broad-based discussion of technological shaping aligned with African values and priorities. Faith institutions are also negotiating their stance towards rising technologies. Some conservative strands resist innovations seen as profaning traditional mores, while progressive groups advocate humane technological adaptation. Most major religious organisations have formulated evolving technology policies balancing pragmatism, ethical cautions and spiritual wisdom.

Indigenous African cosmologies provide philosophical frameworks contextualising both risks and promise. Mbiti's (1969) exposition of the communitarian ethos, 'I am because we are', indicates that technologies resonating with African philosophical values systems can find fertile ground if harnessed for communal benefit rather than self-serving individualism. However, those violating human dignity may face resistance. Thus, conscious localisation negotiating external innovations with local worldviews

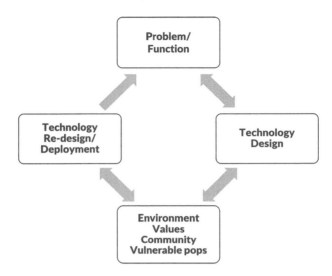

Fig. 6.1   An African social constructivism of technology

remains prudent. At family and community levels, technologies spur mixed outcomes. Mobile phones strengthen family bonds over distance but can distract from present company. Farm mechanisation eases drudgery but ruptures village labour systems if unconsidered. Social media builds connections but fragments if not directed ethically. Therefore, grassroots innovation governance and adaptation respond dynamically to each technology based on lived experiences.

Technologies interfacing with each significant institutional sphere create opportunities for empowerment if guided by communal priorities. However, uncritical adoption risks undermining African institutions if not thoughtfully tailored to complement values. By democratically deliberating on appropriate integration, African communities can selectively adopt and adapt emerging innovations on their terms to serve common aspirations. This social constructivism can be illustrated in Fig. 6.1.

6.4   African Cultures Shaping Technology

African cultures possess rich philosophical and ethical resources that can shape technology's trajectory on the continent in more humanistic, communalistic directions aligned with local values. Rather than technologies disrupting traditions, conscious innovation informed by African cosmologies presents opportunities to elevate technical and social progress.

As Babalola (2017) illustrated, ancient African civilisations like Ile-Ife and Igbo-Ukwu developed sophisticated metalworking, architecture, medicine and governance systems grounded in communal values different from Western individualism. Recuperating such pre-colonial technological wisdom can inspire ethical innovation frameworks resonant with indigenous heritage. Mbiti's (1969) philosophy of integrating the material and spiritual provides philosophical foundations for technology design that respect human dignity and communal ties. Sciences in African cultures were never divorced from the sacred but embedded within holistic worldviews. Early African art integrated science, technology and spirituality rather than separating knowledge (Oladumiye, 2014). This unity of technical arts, social ethics, and metaphysical purpose can steer future innovation trajectories to serve inclusive human development.

Core African philosophical concepts like 'Ubuntu' emphasising communal solidarity over self-interest provide values to ethically direct rising technologies through collective consultation and consent-based adoption. Moral philosophies stressing service, social duties and human relations over profits or efficiency can shape design priorities and regulation principles. As Wiredu (2008) explained, public deliberation in African customary jurisprudence aims at communal consensus. Such ethics translated to technology governance can ensure inclusive decision-making and participatory priority-setting around innovation choices. Rather than technologies dividing communities as individualistic tools, the African customary ethos commends shared technologies like village microgrids that unite community development needs rather than differential private consumption.

Realising such localisation requires going beyond superficial indigenisation, like wearing traditional attire while using Western technologies. Figure 6.1 shows that deeper technological indigenisation entails adapting tools to resonate with African knowledge systems, identities, environments and priorities (Kinyanjui, 1993). For instance, virtual education can localise through weaving in African storytelling and narratives rather than mimicking Western content. Health technologies can respect local healing traditions in syncretic systems. Africapitalism philosophies can guide business innovations to answer communal goals.

African cosmologies grounded in nature and community can shape technologies reflecting ecological balance and social reciprocity. Tech designed around sacred values like life, dignity and collective duty can resist individualistic market logic. As Wiredu (2008) explained, the Akan concept of personhood recognises individual and communal dimensions, providing a framework to harmonise personal and collective aspirations through judicious technological shaping. Reclaiming technologies like maps, writing, and medicines diffused from African origins before colonisation subjugated indigenous knowledge provides a cultural foundation for sovereign innovation (Austen & Headrick, 1983). Revival of pre-colonial technologies on local terms refutes imposed narratives of primitive African techniques compared to borrowed "advanced" tools.

Still, cultures change, and active technological shaping matters more than passive nostalgia. As Mudimbe (1988) warned, valorising pre-colonial Africa risks essentialising cultures when adaptation is constant. Authenticate Africa's diverse heritage while proactively governing rising technologies through anticipated impacts on communal values. Therefore, recognising fluidity alongside wisdom traditions allows for the selective integration of appropriate innovations that empower communities while localising disruptive technologies or restricting those undermining African values (Kinyanjui, 1993). A critical yet pragmatic stance can help realise the promise of emerging innovations to improve human conditions materially and spiritually while mitigating threats to social fabrics if misapplied.

Enterprise, medicine, education, governance, and infrastructure opportunities become possible with technologies aligned to African communal priorities and identities. However, the uncritical adoption of individualistic tools risks exacerbating inequalities, loss of social cohesion and moral drift. A middle path must acknowledge changing aspirations, demographics, and economic pressures while retaining humanistic orientations. Active technology assessment and regulations will help circumvent risks of dissolving family and community ties, spiritual alienation, cultural disruption, or erosion of freedoms through imposed technologies designed without consent. African innovation policies already navigate this complexity, weighing risks of emerging technologies like artificial intelligence, biotechnology, nanotechnology, and persuasive systems against solutions for local challenges like food security, clean energy, environmental conservation, and youth employment. Multidisciplinary foresight and governance help align innovations with ethical priorities of social harmony.

Still, African societies now interface with global systems and cannot wholly resist external paradigms. Some individualist technologies may provide medical, educational, and entrepreneurial advances that are difficult to refuse. Total exclusion risks missed opportunities, while uncritical adoption enables neo-colonial dependency. Therefore, an empowered 'Africanization' stance engages critically with global forces while centring communal values and strategic interests (Obiagu, 2023). Here, cultural wisdom traditions temper—but do not necessarily reject—appropriate technologies. African cultures offer epistemic resources to ethically shape rising technologies for communal benefit rather than individualistic disruption of social fabrics. By consciously building innovation ecosystems rooted in human dignity, social reciprocity, and ecological harmony, African societies can thoughtfully integrate helpful tools and systems while localising or restricting innovations that violate core values. The promise lies in blending science and technology with communal moral wisdom to elevate both.

6.5   The Co-construction Process

Africa's full participation in shaping emerging technologies is essential for innovations to resonate with local realities, priorities, and values. Beyond passively consuming external tools, active African voices and perspectives must inform co-creation processes. Opportunities exist to exhilarate new platforms, content, services, and systems design. Innovations flourish through open exchange of diverse ideas, skills, and resources. Africa possesses rich reservoirs of tacit knowledge, creative talent, and unmet needs to catalyse novel solutions. Providing platforms for user communities, technologists, and cross-disciplinary experts to associate and collaborate freely can drive homegrown innovation tailored for African contexts. Examples like Afrotech conferences, AfriLabs hubs, and Africa4Tech forums that connect developers, entrepreneurs, and social innovators across countries reveal the promise of creative synergy through exchanges. Pan-African technical networks allow the sharing of solutions to common challenges. Joint creativity further emerges from Africa, connecting with diaspora expertise. Co-creating technologies drawing insights from global interactions while preserving local control over innovation shaping aligned with African priorities holds potential for technologies advancing social and business purposes.

Users co-construct technologies through active negotiation during design processes rather than being passive recipients of finished products. Their lived experiences should shape specifications and features. For instance, localisation requires a deep understanding of how target users navigate daily challenges and co-design appropriate solutions leveraging their environment and socio-cultural context. Inclusive innovation platforms recognise that marginalised user perspectives are essential to overcome elite bias. Approaches like Participatory Technology Development bring women, youth, rural farmers and informal workers into collaborative design through the joint definition of problems and solution criteria. Negotiated priorities balancing user aspirations, technical capabilities, business models, and social impacts allow co-constructing technologies to resonate with African settings. Through extensive participatory engagements spanning problem definition, requirement analysis, prototyping, pilot-testing and user feedback, Africans can guide design based on negotiated priorities rather than importing ready-made technical systems.

User participation in technology development contests dominant paradigms imposed through external tools by reclaiming agency to shape innovations on their terms. For instance, civic advocacy groups in Africa should be contesting biased algorithms and machine learning datasets that fail to represent African contexts. Citizen oversight bodies auditing technologies for social impacts provide channels for public critique if innovations undermine democratic values or human rights. Africa's experience countering unjust colonial and postcolonial technologies gives moral authority to raise ethical objections regarding contemporary platforms enabling surveillance, automation-based displacement, or opaque artificial intelligence. Thus, proactive technology assessment contested by empowered users compels creators to align innovations with African values like transparency, accountability, and consensus orientation. User protections like data sovereignty, design safeguards, and regulatory regimes can emerge through such democratic oversight.

As in Fig. 6.1, localisation enables the user to customise technologies to resonate with indigenous cultures, environments, and knowledge systems. The agency's ability to interpret, adapt, and reconfigure tools for contextual fit allows African communities to integrate external innovations on their own terms. For instance, digital learning systems can localise by using animations, vernacular languages, and familiar narratives rather than importing alien paradigms. Agricultural apps guide farmers in using localised crop varieties, weather insights, and community knowledge sharing rather than generic analytics. Offline data access and low-bandwidth interfaces recognise infrastructure constraints. Voice user interfaces aid illiterate users. Such adaptations to match user capabilities and preferences illustrate localisation enabling user agency. User freedom over deciding appropriate integration with local contexts, rather than forced adoption of standardised tools, allows technologies to strengthen rather than disrupt social fabrics and wisdom traditions. Culturally resonant innovation emerges from respecting user interpretation grounded in lived experiences. Through such localisation, external tools become indigenised on African terms.

The spread of social media across Africa demonstrates the double-edged potential of platforms shaped variously by different user communities. In some contexts, social media has become a tool for dangerous disinformation, hate speech, and political manipulation. However, its innate malleability also allows more positive localised adaptation. For instance, youth-driven anti-corruption campaigns leverage social media for civic activism and government accountability. Health campaigns utilise platforms to spread awareness and counter stigma. Religious communities build fellowship and share sermons through messaging apps. Rural entrepreneurs access markets. Such user-driven reinterpretation for public purposes shows the opportunity in local agencies over technological shaping, not just design intentions.

6.6   Augmenting African Voices and Perspectives

Africa's vast diversity and plurality means single narratives cannot adequately represent technology's impacts and promise across varied contexts. Amplifying plural African voices and perspectives through inclusive public engagement and debate provides a more nuanced, grounded understanding of multidimensional socio-technical dynamics frequently obscured in policy conversations dominated by external experts and institutions. Constructivist communication methodologies that respectfully engage diverse African communities can reveal insightful encounters, critiques, aspirations, and ideas that diverge from conventional institutional framings of technology as either universally liberating or oppressively homogenising. More possibilities emerge for ethical innovation shaping that aligns tools and systems with local values through elevating plural voices.

Communicative constructivism entails open-ended public dialogues, digital storytelling, focus groups, participatory media production, and other qualitative techniques that elicit authentic voices reflecting diverse technological experiences and perspectives grounded in lived realities across Africa's heterogeneity (Rodny-Gumede, 2017). The aim is pluralistic perspective sharing without a predetermined agenda or constraining dichotomies that reduce complex dynamics to utopian versus dystopian extremes. Inclusive processes make audible technological encounters of social groups frequently muted in formal policy conversations, like informal workers, remote villagers, women, and youth navigating daily challenges using innovations. Their narrated stories, critiques, aspirations, and ideas often diverge from institutionalised framings of technology. For instance, as Rodny-Gumede (2017) found, digital storytelling platforms for township residents to constructively discuss their encounters with HIV/AIDS treatment and prevention programmes foregrounded localised meanings and community insider perspectives beyond sensationalist media portrayals or official health promotion rhetoric. The nuance revealed through their voiced experiences contested both utopian and dystopian extremes.

Likewise, feminist participatory media and digital platforms that enable women to articulate gender dimensions of technology adoption safely, often made invisible in male-dominated public forums, provide critical counters to dominant narratives. Creating discursive spaces for women to share about barriers, opportunities, achievements, and ideas exposes patriarchal constraints around access, skills, and cultural norms that commonly marginalise female voices. Youth digital expression through music, arts, social media, and blogging reveals generational perspectives on technology's risks and possibilities that frequently differ from elder stances. Such illustration of youth encounters, aspirations, values, and critiques beyond stereotypes enriches intergenerational understanding and policy debates. Communicative constructivism prevents the monolithic framing of technology as either universally beneficial or threatening to African communities. More nuanced possibilities emerge through narrated experiences for ethical innovation shaping that aligns tools with cultural values and social visions. This contracts techno-deterministic assumptions of technologies inherently leading to pre-defined outcomes without considering pluralistic shaping.

Indeed, qualitative insights from communicative engagement with diverse publics have a growing influence on technology policymaking compared to quantitative analytics. The Millennium Development Goals preceded the Sustainable Development Goals in advancing multilateral frameworks for international development that combine expert indicators with participatory processes, engaging communities as stakeholders, not just data points. Likewise, government interventions around expanding

Internet access, digital skills training, e-government services, data protection, automation planning and other areas increasingly open consultative forums for plural voices to inform policy priorities and designs rather than relying entirely on technical bureaucracies. Constructivist listening provides democratic grounds for achieving contextual relevance and social legitimacy amid technology disruptions.

While communicative constructivism reveals multifaceted social perspectives on technology, critical constructivism examines deeper power hierarchies shaping whose voices and priorities dominate institutional technology discourses while others get marginalised (Adams, 2006). It applies critical theory to interrogate force relations determining meaning. Critical analysis in the tradition of Foucault exposes how governments, corporations, and institutions exert epistemic hegemony, constructing influential ideologies and discourses that enshrine particular interests at the expense of more marginalised populations who pay the hidden costs. Specific versions of Africa's technology narrative gain hegemonic dominance while others remain subjugated.

African critical constructivism must confront whose social and economic interests are served by normalised technocratic paradigms like AI and biotechnology that promise progress but potentially displace and disempower communities. It must challenge façades of supposed technological neutrality that obscure biases and contest zones of silence where failures and unintended harms of celebrated innovations get buried through propaganda amplifying selective success stories that maintain mythologies of linear progress and win-win advancement. For instance, popular social media platforms project inclusive narratives of democratised opportunity, flexible entrepreneurship, and empowerment through technology access. However, poor rural communities are frequently left behind due to digital divides, women excluded by harassment and skills gaps, informal workers losing protections, and cultural erosion. Such populations experience the costs while elite demographics disproportionately gain. Likewise, policy visions in many African countries celebrate the Fourth Industrial Revolution technologies like artificial intelligence for growth and leadership and sound compelling on the surface. However, erosion risks mass employment from premature automation in economies with limited social safety nets.

Further, African critical constructivism must compel more holistic consideration of the varied human impacts of technologies, not just their efficiency, by interrogating the exercise of institutional power shaping discourses. It must call for structural reforms that empower African communities to more democratically determine technological trajectories so tools explicitly serve society's interests, not just status quo power structures. Otherwise, without such epistemic justice, people risk becoming objects acted upon by technologies of control determined by others rather than gaining recognition as empowered subjects who can shape innovations, as many African scholars warn, drawing on philosophical traditions of communal ethics. Technologies devoid of participatory shaping can undermine human dignity.

The point is not outright technology rejection but rather inclusive participatory agency over design directions. Critical constructivism provides analytical leverage for marginalised populations impacted by technologies to contest injustices, articulate concerns, and shape tools to align with their values on their terms. This contracts externalist shaping dominated by elite interests that ignore the lived realities of ordinary citizens who pay the price of disruption. Ultimately, both communicative and critical constructivist lenses reveal greater complexity at the intersection of society and technology than prevalent policy discourses focused on technical metrics, passive adoption and assumed linear progress. Constructivism foregrounds people's active meaning-making and diverse experiential encounters with socio-technical change. It empowers plural voices in navigating technology's opportunities and risks in ethical directions.

Genuinely inclusive participation in shaping Africa's technology discourses and policies requires constructively engaging voices of key social constituencies frequently excluded or sidelined from dominant institutional conversations: Rural communities provide counternarratives that reveal limitations of technologies designed primarily for urban users, as well as infrastructural barriers to digital access faced by rural citizens. Their place-based perspectives on livelihood and cultural impacts are indispensable for holistic innovation policymaking. Further, indigenous communities articulate unique experiences of cultural impacts from technologies violating traditional lifestyles. Their participation provides ethical grounding. While more senior communities may lack youth digital fluency, their wisdom and experience offer prudent guidance on aligning technologies with African values threatened by the uncritical adoption of external tools. Their oversight role remains vital.

Constructively engaging the entire plurality of African societies is indispensable for emerging technologies to advance inclusive development rather than exacerbating inequalities and disruption. Expanding discursive spaces allows marginalised populations to articulate local realities, contest biased framings, and shape tools meeting their needs on their terms. The resulting diversity of African voices will convey technology's multifaceted social meanings and possibilities more richly than limited technocratic framings. With inclusive participation, Africa can lead in centring values, not just technical metrics, in navigating the planet's unfolding era of technological transformations.

6.7   Conclusion

This chapter has attempted to develop social constructivist perspectives as conceptual lenses to examine technology innovation in Africa. In contrast to technological determinism, constructivism recognises technologies as socially shaped artefacts embedded within cultural contexts. Key arguments demonstrate technologies arising from agendas and interests, problematise assumed neutrality, articulate possibilities for localisation on African terms, and advocate elevating marginalised voices in shaping trajectories. The aim is to establish foundations for situated, humanistic African innovation guided by local values. Key insights include recognising innovation as a process of multidirectional shaping between technologies and societies. Neither unfolds unilaterally but through reciprocal interactions, negotiations, and contests reflecting diverse priorities. Making visible the underlying interests, assumptions, and agendas behind technological paradigms is indispensable for more equitable co-construction of African innovation futures that empower marginalised groups.

This conceptual foundation opens vital space for situated, humanistic African innovation pathways firmly rooted in local epistemologies, cosmologies, and participatory shaping. The outlook encompasses technologies guided by cultural philosophies of ethics and solidarity to improve human conditions materially and equitably. It envisions participatory processes engaging plural voices to direct tools serving African visions of dignified, holistic development on society's terms. Moreover, it entails mindful localisation and assessment, continually aligning external systems with indigenous values. Overall, social constructivism provides indispensable guidance for governance, enabling Africa's unfolding technological opportunities to be directed democratically towards human priorities through insider participation, not unilaterally imposed from outside. This chapter has sought to establish conceptual foundations for this hope of ethical, situated innovation across Africa's diversity.

REFERENCES

Adams, P. (2006) Exploring social constructivism: Theories and practicalities. Education, 3–13, *34* (3). pp. 243–257. https://doi.org/10.1080/03004270600898893.

Austen, R. A., & Headrick, D. (1983). The role of technology in the African past. *African Studies Review, 26*(3/4), 163–184. https://doi.org/10.2307/524168

Babalola, A. B. (2017). Ancient history of technology in West Africa: The indigenous glass/glass bead industry and the society in early Ile-Ife, Southwest Nigeria. *Journal of Black Studies, 48*(5), 501–527. http://www.jstor.org/stable/44631296

Bick, R., Halsey, E., & Ekenga, C. C. (2018). The global environmental injustice of fast fashion. *Environmental Health, 17,* 92. https://doi.org/10.1186/s12940-018-0433-7

Bijker, W. (1995). Sociohistorical technology studies. In S. Hasanoff, G. E. Markle, J. C. Peterson, & T. J. Pinch (Eds.), *Handbook of science and technology studies* (pp. 229–256). Sage.

Bijker, W. E., Hughes, T. P., & Pinch, T. (1987). General introduction. In W. E. Bijker, T. P. Hughes, & T. J. Pinch (Eds.), *The social construction of technological systems: New directions in the sociology of history and technology* (pp. 1–6). MIT Press.

Bijker, W. E., & Law, J. (1992). Do technologies have trajectories? In W. E. Bijker & J. Law (Eds.), *Shaping technology/building society: Studies in socio-technical change* (pp. 17–20). MIT Press.

Ellul, J. (1964). *The technological society, vintage books* (J. Wilkinson Trans.). Random House.

Fairchild, A. M., & Quansah, E. A. (2007). Approaching the digital divide in Sub-Saharan Africa: Technological determinism or social constructivism? *International Journal of Knowledge and Learning, 3,* 612–627.

Feenberg, A. (1991). *Critical theory of technology.* Oxford University Press.

Glover, D., Sumberg, J., & Andersson, J. A. (2016). The adoption problem; or why we still understand so little about technological change in African agriculture. *Outlook on Agriculture.* https://doi.org/10.5367/oa.2016.0235

Gyekye, K. (1997). *Tradition and modernity: Philosophical reflections on the African experience.* Oxford University Press.

Hickman, L. A. (1992). *John Dewey's pragmatic technology.* Indiana University Press.

Hughes, T. P. (1987). The evolution of large technological systems. In W. E. Bijker & T. P. Hughes (Eds.), *The social construction of technological systems* (pp. 51–82). MIT Press.

Kinyanjui, K. (1993). Culture, technology and sustainable development in Africa. *Asian Perspective, 17*(2), 269–295. http://www.jstor.org/stable/42704030

Law, J. (1987). Technology and heterogeneous engineering: The case of Portuguese expansion. In W. E. Bijker, T. P. Hughes, & T. J. Pinch (Eds.), *The social construction of technological systems: New directions in the sociology and history of technology* (pp. 111–134). MIT Press.

MacKenzie, D., & Wajcman, J. (Eds.). (1985). *The social shaping of technology*. Open University Press.

Mbiti, J. S. (1969). *African religions and philosophy*. Heinemann.

Mcquarie, D., & Amburgey, T. (1978). Marx and modern systems theory. *Social Science Quarterly, 59*(1), 3–19. http://www.jstor.org/stable/42859863

Moon, N. W., Baker, P. M., & Goughnour, K. (2019). Designing wearable technologies for users with disabilities: Accessibility, usability, and connectivity factors. *Journal of Rehabilitation and Assistive Technologies Engineering, 6*. https://doi.org/10.1177/2055668319862137

Mudimbe, V. Y. (1988). *The invention of Africa: Gnosis, philosophy, and the order of knowledge*. James Currey.

Mumford, L. (1934). *Technics and civilization*. Routledge.

Murove, M. F. (Ed.). (2009). *African ethics: An anthology for comparative and applied ethics*. University of KwaZulu-Natal Press.

Ndasauka, Y. (2021). Dynamic view of technology: Implications on ethics of social networking sites. In G. Kayange & C. Verharen (Eds.), *Ethics in Malawi*. RVP.

Obiagu, A. N. (2023). Toward a decolonized moral education for social justice in Africa. *Journal of Black Studies*. https://doi.org/10.1177/00219347231157739

Oladumiye, B. E. (2014). Perception of relationship between art and science in contemporary African arts and technology. *International Journal of Art, Culture, Design, and Technology (IJACDT), 4*(2), 51–63. https://doi.org/10.4018/IJACDT.201407010

Pannabecker, J. R. (1991). Technological impacts and determinism in technology education: Alternate metaphors from social constructivism. *Journal of Technology Education, 3*(1), 1–11. https://doi.org/10.21061/jte.v3i1.a.4

Pesando, L. M., & Rotondi, V. (2020). Mobile technology and gender equality. In W. Leal Filho, A. Azul, L. Brandli, A. Lange Salvia, & T. Wall (Eds.), *Gender equality. Encyclopedia of the UN sustainable development goals*. Springer. https://doi.org/10.1007/978-3-319-70060-1_140-1

Pinch, T. J., & Bijker, W. E. (1984). The social construction of facts and artefacts: Or how the sociology of science and the sociology of technology might benefit each other. *Social Studies of Science, 14*(3), 399–441. https://doi.org/10.1177/030631284014003004

Porter, G., Hampshire, K., Milner, J., Munthali, A., Robson, E., de Lannoy, A., Bango, A., Gunguluza, N., Mashiri, M., Tanle, A., & Abane, A. (2015). Mobile phones and education in Sub-Saharan Africa: From youth practice to public policy. *Journal for International Development*. https://doi.org/10.1002/jid.3116

Rodny-Gumede, Y. (2017). Questioning the media and democracy relationship: The case of South Africa. *Communications, 43*(2), 10–22. https://doi.org/1 0.1080/02500167.2017.1318938

Willoughby, K. W. (1990). *Technology choice: A critique of the appropriate technology movement.* Westview Press.

Winner, L. (1980). Do artifacts have politics? *Daedalus, 109,* 1.

Wiredu, K. (2008). Social philosophy in postcolonial Africa: Some preliminaries concerning communalism and communitarianism. *South African Journal of Philosophy, 27*(4), 332–339.

CHAPTER 7

Harmonising Technology and Humanity Through Ubuntu

7.1 Introduction

The ethical foundations for technology development and integration in Africa are in need of rethinking. Mainstream innovation paradigms, shaped by Western individualism, neoliberal economics, and instrumental rationality, starkly contrast indigenous African values of communal bonds, social welfare, and ecological harmony (Metz, 2007). In this context, Ubuntu philosophy, rooted in sub-Saharan cultures, offers a humanistic worldview that can transform dominant assumptions and guide technology for collective advancement. As a relational theory of moral ethics, Ubuntu emphasises that our shared humanity is realised through communal relationships, mutual caring, and cooperative social fabrics rather than atomistic self-interest. This perspective challenges Western assumptions of individualism that permeate technology models. By upholding dignity and social justice as innovation transforms African societies, Ubuntu provides valuable wisdom for navigating these changes. Reviving Africa's philosophical foundations is crucial to counteract the uncritical transfer of foreign technical systems and paradigms that fuel dependence, inequality, and loss of identity (Naude, 2019). Grounded in local knowledge, Ubuntu offers a framework for directing development and technology integration based on indigenous values and visions. As Africa experiences rapid sociotechnical change, Ubuntu thought inspires new perspectives across diverse fields, from business ethics to human rights, by communicating principles

Y. Ndasauka, *African Mind, Culture, and Technology*, https://doi.org/10.1007/978-3-031-62979-2_7

of solidarity and consensual governance that resonate through traditional cultures (Kayange, 2018).

Mainstream globalised technology regimes, often oriented towards productivity, efficiency, and control, frequently fail to account for negative social consequences in Africa. Further, perspectives shaped by cost-benefit calculations rather than moral reasoning risk exacerbating inequality, exploitation, loss of meaning, and cultural disruption across the region. In contrast, Ubuntu's communitarian lens provides a philosophical grounding to critically assess technologies and steer innovation trajectories in ways that nourish collective interests and social welfare, not just growth. Its principles of solidarity, justice, and sustainability should guide technology integration. Moreover, Ubuntu underscores that genuine progress requires African innovation pathways to be grounded in local knowledge systems and self-determined visions rather than the unreflective import of foreign paradigms. Technocracy must be tempered with traditions, and centring community agency and wisdom in shaping appropriate, humanistic technologies is imperative.

This chapter argues that Ubuntu philosophy enables the assessment and steering of technologies to nurture social cohesion and avoid harm. Its principles, such as solidarity and care for the vulnerable, guide ethical and empowering innovation. Ubuntu emphasises that technology integration must be grounded in indigenous knowledge and locally defined visions of progress, with community agency and wisdom playing a vital role in developing appropriate and sustainable solutions. The chapter begins by outlining Ubuntu ethics' communal values and social justice foundations. It then discusses the potential of Ubuntu thought in evaluating technologies and modelling endogenous innovation responsibly and critically. Case studies are presented to assess the impacts of agricultural and health technologies on communities through an Ubuntu lens. Finally, the chapter proposes strategies for applying Ubuntu principles in shaping technologies that uphold human dignity across Africa's diverse locales, demonstrating the importance of Ubuntu in developing contextual technology ethics.

7.2   Ubuntu as an Ethical Theory

Ubuntu is an ethical and philosophical theory deeply rooted in sub-Saharan Africa's indigenous cultures and communal values. The term 'Ubuntu' emerges directly from the Bantu languages of Southern Africa, including isiXhosa, isiZulu, Runyoro, and Luganda, and it is most commonly translated into English as expressing the concepts of 'humanness' or 'humanity'. However, the meaning of Ubuntu is richer and more complex than any single English word can fully convey. At its core, Ubuntu ethics emphasises the interconnectedness of human existence and the importance of relationships, mutual caring, and social harmony within a community (Metz, 2007). As an ethical framework for how to live a good life, Ubuntu upholds ideals of compassion, solidarity, cooperation, and collective responsibility. It views human persons not as isolated, atomistic individuals but rather as fundamentally shaped through their bonds and interactions with other people within a society (Ramose, 1999; Metz, 2011). The much-quoted Zulu maxim 'umuntu ngumuntu ngabantu' vividly encapsulates this relational ontology central to Ubuntu thought, meaning, 'a person is a person through other people'. In the Ubuntu worldview, our full human potential can only be realised through fellowship with others, not in isolation. Ubuntu reminds us that social interconnection and interdependence are inescapable dimensions of our existence.

In addition to recognising human interconnection, Ubuntu also firmly upholds every person's inherent dignity and fundamental equality. This underpins its egalitarian values of social justice, mutual caring, and reciprocal obligations to support each other's well-being, which are found in Ubuntu (Metz, 2011). Since we all exist through community, the welfare of each individual is intertwined with the welfare of society as a whole. The Ubuntu theory of ethics prescribes that human beings all have ethical responsibilities to employ their capabilities in service of others' development, to exhibit virtues like generosity, kindness, compassion, honesty, and empathy in their relationships and social dealings, as well as to respect the dignity of every person regardless of their social standing or background. A spirit of humaneness towards others, especially the vulnerable, resonates through the Ubuntu ethical perspective. Overall, Ubuntu provides a humanistic moral outlook for African cultures centred on the values of communal existence, solidarity, cooperation, and compassion.

Some scholars note that the ethical guidance of Ubuntu is directed towards both interpersonal dealings and broader applicability for leadership, governance, and policy in society. For instance, political leaders and governments guided by Ubuntu ideals are expected to consult widely with citizens, make decisions through participatory consensus-building, and share resources equitably—rather than exercising power in a self-interested or authoritarian manner (Metz, 2014). Those in authority carry responsibilities as public servants to enact the will of local communities for the common good. Reconciliation of conflicts and differences through open dialogue is vital. Overall, the principles of Ubuntu aim to nurture social harmony through the joint pursuit of shared goals. While originating from sub-Saharan indigenous cultures, the humanistic essence of Ubuntu continues to inspire evolving interpretations and applications as a moral framework for addressing contemporary ethical dilemmas in fields as diverse as business, education, social policy and technology (Kayange, 2018). Its emphasis on human dignity, social justice, ecological sustainability, and democratic consensus retains deep value in modern contexts, even as some aspects require thoughtful updating. Ubuntu offers wisdom to revive traditional African spiritual ethics on collective solidarity and mutual care in a way that resonates with current challenges.

A foundational component of the Ubuntu ethical perspective is its strong emphasis on communitarian values that privilege social interdependence, collective interests and group solidarity over individualism (Metz, 2007). In the Ubuntu philosophy, human identity and agency are fundamentally constituted by one's participation and relationships within a community rather than existing as entirely self-sufficient atoms (Ramose, 1999). This places great ethical weight on social relationships and cooperation as crucial to realising our shared humanity. Specific communal values prioritised in Ubuntu thought include cohesion and belonging within a group, inclusive participation in decisions affecting the collective, reciprocity in caring for each others' welfare, building consensus through dialogue, and loyalty towards the greater community good (Koenane & Olatunji, 2017). The virtues and behaviours considered ethically sound are those directed outward to support others in society, such as exhibiting generosity, hospitality, compassion, kindness, honesty, forgiveness, and a spirit of sharing one's talents and resources for the advancement of all. Our shared bonds and interconnection as human beings are seen as the basis for having reciprocal duties to employ one's capabilities in service of others.

In the political realm, Ubuntu emphasises participatory, inclusive decision-making that integrates diverse viewpoints through open deliberation and consensus-building, as well as equitable distribution of public resources for the benefit of all citizens. Leaders must be accountable to local communities, exercising authority in a spirit of transparency, responsibility, and care for the marginalised. Reconciliation of conflicts through non-violent dialogue and promoting social solidarity and cooperation towards collective goals are vital. Overall, the communal ethics underpinning Ubuntu thought privilege nurture collective well-being, social harmony, and the public good over egoistic individualism.

The ethical framework of Ubuntu firmly upholds principles of social justice and protection of human dignity, derived from its communitarian foundations that recognise our shared interconnection (Ramose, 2002; Metz, 2011). Since human identity and well-being are seen to be constituted by the community, the welfare of each individual is interlinked with the interest of all in the Ubuntu worldview. Oppression or inequities facing any members of society are seen as diminishing the humanity of all. Consequently, respecting human dignity necessitates mutual concern for equitable freedoms, capabilities, resources, and quality of life across a community. Ubuntu theorist Metz (2011) elaborates that its social ethics require fair and inclusive distribution of public resources, political participation in decision-making processes regardless of status, equality of opportunities, and enacting special protections for the marginalised and vulnerable. Since our fates are intertwined, advancing human development is viewed as a collective responsibility in Ubuntu philosophy—the upliftment of each is only possible through the upliftment of all. Where necessary, Ubuntu demands structural and policy reforms to remedy injustices and nurture a more egalitarian society marked by mutual caring.

Additionally, Ubuntu's conception of human dignity calls for everyone to be respected equally regardless of their social standing or background, in contrast to class hierarchies in modernist societies. Traditional sub-Saharan social structures were more horizontal and democratic. Ubuntu's priority for social justice and dignity inspires to remedy ongoing forms of oppression and marginalisation in postcolonial Africa, affirming the equal worth and membership of groups like women, ethnic minorities and lower classes whose personhood has been diminished. Ubuntu centres a robust, holistic concept of social justice, emphasising equitable capabilities, resources, participation, solidarity, and recognition for all people based on inherent human dignity.

Scholarly debates highlight some critiques of Ubuntu ethics, including ambiguity, romanticisation, and Africanization controversies. Some see the need for more elaboration on adjudicating communal and individual rights as a weakness (Matolino & Kwindingwi, 2013). Metz (2014) responds that Ubuntu allows individuals to limit their freedoms to prevent social harm with reasonable burdens. However, guidance is required to balance tensions. Additionally, the idyllic communitarian depictions of traditional African societies mask past oppressions like patriarchy, geron-tocracy, and tribal marginalisation (Matolino & Kwindingwi, 2013). However, reciprocity and social justice remain applicable if interpreted progressively. In addition, scholars debate whether Ubuntu should be treated as a universal ethic or an African particularity (van Niekerk, 2007). Metz (2011) argues that its humanistic underpinnings have some cross-cultural resonance, but Ubuntu arises from and remains rooted in Southern African cultures, retaining this heritage. Integrating other contextual insights arguably strengthens Ubuntu without relinquishing its identity.

While originating from indigenous cultures, evolving modern interpre-tations highlight the continued relevance of Ubuntu as a humanistic ethi-cal framework for guiding policy, education, business, and development initiatives in Africa today. For example, scholars like Kayuni and Tambulasi (2012) have illustrated the potential of Ubuntu virtue ethics to transform corporate decision-making and operations in Africa beyond narrow profit motives towards more holistic objectives that benefit communities and the environment. In education, integrating the communitarian values of empathy, cooperation, and inclusion inherent in Ubuntu is seen as an antidote to excessive individualism among youth (Waghid, 2014). Development programmes are also increasingly guided by Ubuntu's par-ticipatory, socially just principles to enhance community self-reliance and collective advancement. These cases illustrate the ongoing adaptation of Ubuntu's humanistic essence to address ethical challenges across contem-porary spheres, grounding reforms and leadership in African philosophical wisdom. Ubuntu's relational worldview and moral principles offer mod-ern societies a lens to nurture social interdependence, compassion, and dignity. Respectfully integrating indigenous ethics thus retains relevance, even within changed contexts. While some aspects of Ubuntu require thoughtful updating, its core tenets continue providing vital guidance to revive traditional African spiritual values of collective solidarity and mutual care in a way that resonates with current social, economic and environ-mental challenges. Ubuntu's humanistic foundation carries enduring worth as Africa seeks homegrown philosophical perspectives to inform ethical governance, responsible business, and social policy for today's realities.

7.3   Ubuntu and Technology

The communitarian values and social justice principles within the African ethical philosophy of Ubuntu offer a constructive perspective to critically evaluate and guide technology innovation trajectories to align with collective interests. Ubuntu's relational worldview highlights potential harms from technologies that disrupt social bonds and community fabrics in the pursuit of efficiency or profits. For instance, automation and artificial intelligence that sacrifice jobs and workplace relationships for productivity contradict Ubuntu ideals of human dignity and solidarity, even if economic output rises. Labour rights and communal impacts should take priority over capital interests. Equally, digital platforms and social robots replacing caring human interactions undermine Ubuntu's emphasis on ethical interconnection. Virtual connections without empathy violate the deeper meaning of our shared humanity in Ubuntu thought.

Ubuntu justice principles also provide grounds to resist extractive and exploitative production systems that destroy ecosystems and community livelihoods. Economic development achieved through violations of collective interests and sustainability is considered unethical from the Ubuntu perspective. Overall, the rich communitarian ethics and social justice foundations of the Ubuntu philosophy offer invaluable guidance for assessing and directing technology innovation in ways that nourish collective wellbeing and the public good rather than fuelling fragmentation, inequality, and dehumanisation.

The communitarian ethics and social justice foundations of the African philosophy of Ubuntu offer invaluable guidance for shaping more responsible, ethical and endogenous models of technological innovation that uplift and empower African societies based on local values, interests, and developmental visions. Scholars argue that fully realising the potential of Ubuntu thought requires reforming dominant research and business cultures that remain shaped by colonial legacies and neoliberal ideologies to consciously apply Ubuntu principles of collective participation, consensus-building, social justice and ecological sustainability more deeply into innovation ecosystems (Gwagwa et al., 2022; Naude, 2019). Technology development guided by an Ubuntu ethical framework would prioritise designing solutions that increase self-reliance and democratise opportunities for marginalised groups rather than exacerbating inequalities and dependence on the West through thoughtless technical interventions. To enable such appropriate, socially conscious innovation, building local skills, knowledge, institutional capabilities, and resource bases is imperative, rather than automatically importing technical systems and solutions from abroad. Education, research policies, and community programmes must actively nurture African creativity, know-how and agency to guide technology development trajectories in ways that serve the interests and priorities of African societies on their terms. Integrating external insights can assist progress if respectfully tailored to local contexts but should only dominate some processes.

Additionally, participatory design practices inspired by the communal values of Ubuntu could facilitate more inclusive and ethical technology development that holistically resonates with African realities. Regular constructive engagements between technology designers, policymakers, researchers, community leaders and ordinary citizens through consensus-oriented forums are essential to maximising social benefits and responsible technology governance. The wealthy philosophical foundations of Ubuntu ethics can provide guidance to transform prevailing research and business ecosystems in African countries towards models of responsible innovation that consciously apply indigenous African wisdom rather than blind technocracy. Ubuntu's humanistic principles and relational worldview necessitate centring local agency, priorities and values to steer technology trajectories in ways that enable collective upliftment and advancement.

The humanistic essence and deep ethical values inherent to the African philosophy of Ubuntu provide vital guidance for consciously shaping technology design, assessment and integration in more socially empowering directions that benefit all of Africa's diverse communities. First and foremost, technologies should be oriented to nurture human bonds and social cohesion rather than undermine communal ties. Responsible automation must retain dignified, meaningful work, while digital tools should primarily enrich real-world relationships, not supplant them (see Friedman, 2023). The relational lens of Ubuntu reveals dehumanising externalities from technologies fixated on efficiency and control rather than solidarity. Second, the Ubuntu ethic of compassion morally obliges innovations to expand capabilities and opportunities, especially for vulnerable populations, through assistive, localised and accessible design (Amugongo et al., 2023). Any technologies violating fundamental human rights or persons' dignity would fundamentally contradict the Ubuntu philosophy's core tenets. Third, respecting the autonomy and worth of all human beings means ensuring technologies empower communities to shape their futures rather than dominate through embedded biases. Algorithmic systems reflecting income, gender, racial or other prejudices offend the universal human value that Ubuntu recognises in all people (Gwagwa et al., 2022). Instead, innovations guided by Ubuntu principles should aim to increase self-determination.

Additionally, social justice is a pivotal Ubuntu value demanding innovations democratise economic and social opportunities, especially for marginalised groups, rather than confer privileges disproportionately to the powerful and further engrain inequality. For instance, responsible implementation of emerging technologies like AI in fields such as healthcare guided by an Ubuntu ethical framework would strive to equitably expand access, quality and affordability for all Africans rather than principally benefit urban elites (Amugongo et al., 2023). Further, the Ubuntu notion of human interconnection with nature requires technology production to safeguard ecological sustainability rather than fuel blind extraction and materialism. Just transitions to circular economies guided by Ubuntu's communal ethics are imperative. Consciously applying the humanistic essence of Ubuntu ethics allows directing innovations to nurture collective upliftment, social equity and sustainability across Africa's diverse communities. This necessitates centring the critical values of preserving human bonds, demonstrating compassion, protecting dignity, advancing justice and balancing material progress with ecological harmony. An Ubuntu lens reveals how technology is shaped for humanity, not just profit, and holds liberating potential.

Modern agricultural technologies, including improved seeds, irrigation systems, farm machinery and agrochemical inputs, are often promoted as necessary solutions to raise productivity and address hunger and food insecurity challenges across Africa. However, analysing the impacts of these technologies through the ethical lens of Ubuntu philosophy reveals a more nuanced picture. While modernisation inputs can deliver higher per-acre yields in optimal conditions, some have decreased overall food production and community resilience by displacing better-adapted traditional crop varieties and farming techniques optimised over generations for local soils, rainfall patterns and community preferences (Zimunya et al., 2015). Introducing hybrid seeds dependent on seasonal repurchase rather than allowing saving and exchanging hardy local seed varieties often cultivated for centuries undermines the food self-reliance principles of Ubuntu. Mechanisation like tractors also frequently benefits larger industrialised farms more than smallholder peasants, resulting in the concentration of productive farmlands and displacement of rural labour. Widespread use of synthetic chemicals has also damaged long-term soil fertility, biodiversity and community health, contradicting Ubuntu's emphasis on ecological sustainability and collective well-being. The erosion of communal farming livelihoods and dignity through such disruptive transitions violates the cooperative values and social harmony in Ubuntu philosophy.

Thus, while simplistic productivity statistics may improve with standardised agricultural technologies, assessing the impacts holistically through an Ubuntu ethics lens reveals trade-offs affecting local capabilities, adaptiveness, autonomy, health, social fabrics, livelihood resilience, ecological integrity and indigenous knowledge systems accumulated over generations. An Ubuntu approach would necessitate adopting agricultural solutions consciously to promote collective interests by strengthening communal agricultural systems and capabilities, sharing innovations inclusively to expand opportunities equitably, including for women and youth, designing approaches building on traditional knowledge rather than displacing it, and assessing impacts across interconnected social, economic, and environmental dimensions relevant to local communities. There are no easy technical fixes—solutions must be grounded in ethical principles, cultural values, and indigenous wisdom. Overall, analysing agricultural technology impacts through an Ubuntu ethical perspective underscores the need for shifts to participatory, socially conscious innovation pathways that put communal needs, capabilities, and ecological sustainability ahead of technocratic productivity paradigms. This requires aligning research, business, and policy ecosystems with Ubuntu's humanistic priorities of collective advancement and social justice rather than individualistic motives. The rich ethics of Ubuntu are thought to provide vital guidance to direct more ethical and appropriate agricultural futures in Africa.

Assessing the impacts of modern agricultural technologies on rural livelihoods and dignity in Africa reveals mixed outcomes that require thoughtful, ethical analysis through the lens of Ubuntu philosophy's communitarian values and social justice principles. Agricultural modernisation promises to raise smallholder incomes by integrating them into commercial farming and global value chains. However, realities on the ground often prove more complex. Input-intensive monocropping methods frequently burden struggling peasant farmers with unaffordable seeds, fertilisers and equipment costs, plunging them into debt traps and dependence on distant corporations (Zimunya et al., 2015). Promises of higher productivity also rely on irrigation infrastructure, soil fertility and favourable weather conditions absent for many rural farmers.

Pushing cash-cropping for exports guided by market incentives rather than communal food needs often undermines the resilience of diverse local crops and culinary traditions, violating principles of self-reliance and ecological sustainability within Ubuntu ethics. Meanwhile, post-harvest agro-processing and trading become concentrated in the hands of exploitative intermediaries, leaving primary producers impoverished. At the same time, automation of plantation-style farming and food manufacturing displaces agricultural employment and traditional livelihoods, even if new jobs eventually emerge higher up supply chains, as workers face transitional hardships (Naude, 2019). Thus, while thoughtfully designed and disseminated appropriate technologies can assist rural livelihoods, the Ubuntu perspective highlights risks of traditional smallholder displacement, rising inequality, indebtedness, loss of capabilities and dignity, and subjection to volatile global markets through careless modernisation drives lacking humanistic foundations. Truly advancing rural communities requires participatory processes to develop agricultural solutions from the bottom up in ways that protect communal interests, capabilities, and knowledge systems. It also summons compassion and social responsibility from policymakers and businesses towards peasants who are often treated as disposable by elite-driven agendas.

Emerging biotechnologies involving genetic modification, gene editing, tissue culture and synthetic biology are often promoted as likely solutions to raise agricultural productivity and address food challenges in Africa but raise several ethical concerns from the perspective of Ubuntu philosophy. Proponents argue potential benefits like enhancing pest resistance, drought tolerance, and nutritional quality of staple crops (e.g. golden rice). However, the risks of disrupting local ecosystems, biodiversity, and sustainable agroecological practices by introducing lab-engineered organisms violate the environmental custodianship principles of Ubuntu ethics. Furthermore, patenting lifeforms to treat nature as capitalist commodities counters the Ubuntu conception of shared communal trusteeship and custodianship of living ecosystems.

Additionally concerning are unethical practices like transgenic animal research, which instrumentalise life in ways that offend dignity. Patents on genetically altered seeds also often prohibit traditional smallholder open seed saving and sharing practices that sustain collective resilience (Peschard & Randeria, 2020). Overall, introducing powerful new biotechnologies must be carefully scrutinised against the benchmark of ethical criteria centred on collective well-being and sustainability. The precautionary principle warrants slowing the implementation of technologies with uncertain and irreversible consequences. Biotech corporates and developers should respect the perspective Ubuntu philosophy provides on balancing productivity ambitions with social justice and ecological integrity.

Applying an Ubuthu ethical highlights problematic tendencies of excessive transfer of foreign agricultural technologies and turnkey solutions into Africa without adequate customisation for smallholder contexts or ecological alignment (Zimunya et al., 2015). The normative assumptions of technical universalism and hubristic paradigms of 'modernisation' reflected directly importing foreign innovations often disrespect local realities. The imposition of exogenous technologies violates the Ubuntu principle that solutions should be grounded in indigenous priorities, knowledge systems, and participation. Turnkey projects delivered from abroad frequently benefit recipient communities little, failing to enhance capabilities. However, more participatory design approaches that respect farmer perspectives, synthesise external insights locally through trial and adaptation, and democratise technology choices can better uphold ethics of empowerment, autonomy and ecological sustainability embodied in Ubuntu thought (Zimunya et al., 2015).

Additionally, innovation exchanges between communities in the Global South with comparable contexts promise to share appropriate solutions guided by Ubuntu solidarity. Ultimately, agricultural technology pathways must align innovation with communal needs, dignity, self-determination, and ecological integrity to enable collective advancement in African terms rather than through the disruptive imposition of inconsistent external models. The wisdom of Ubuntu philosophy provides vital guidance to reform dominant paradigms and put African farmers and communities at the centre of shaping more ethical agricultural futures.

Figure 7.1 summarises the key aspects of applying Ubuntu values to technology, emphasising the importance of nurturing human bonds, expanding capabilities, respecting autonomy, promoting social justice, ensuring ecological sustainability, and centring local agency and priorities. By consciously incorporating these principles into technology development and implementation, African societies can shape innovation trajectories that align with their collective interests, cultural values, and developmental visions.

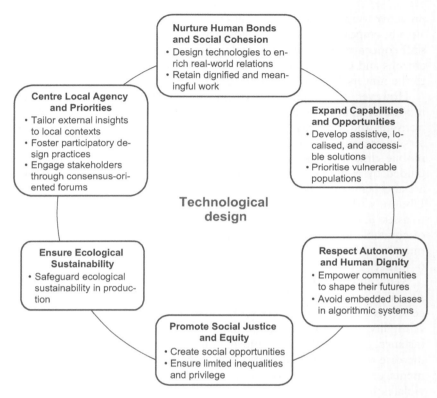

Fig. 7.1 Ubuntu values applied to technology

7.4 HEALTH TECHNOLOGY AND UBUNTU ETHICS

Assessing health technologies through the ethical lens of Ubuntu philosophy spotlights the imperative of evaluating impacts on equitable access and social inclusion in healthcare. While innovations like telemedicine often aim to expand service reach, the benefits may be concentrated among privileged populations if adoption gaps widen along income, geographic, or literacy divides—violating Ubuntu's social justice tenets (Amugongo et al., 2023). For instance, many advanced digital health tools require smartphones, reliable Internet and specialised tech proficiency, limiting their utility for lower-income, rural and elderly groups (Lyles et al., 2023). An Ubuntu perspective presses for intentionally inclusive design and proactive outreach to serve marginalised communities, not just those already empowered. Similarly, advanced medical equipment and quality staff concentrate in urban hospitals, physically excluding rural and poorer citizens and entrenching inequitable dual-track health systems—contrary to the universal care principles of Ubuntu.

However, an Ubuntu orientation can also inspire more participatory models of healthcare innovation that leverage traditional communal relationships and social capital to democratise access to quality care in underserved regions. For example, community health worker programmes, mobile clinics, and popular health education can exemplify innovations guided by communitarian values (see Fig. 7.1). Telemedicine platforms are also being adapted to illiterate users through voice interfaces. Overall, Ubuntu's ethics of care, empathy, and social inclusion highlight the need to assess and steer health technology adoption pathways through the lens of equitable access and capabilities for all. Innovation designed for communal well-being over profiteering is critical to upholding Ubuntu's vision of justice.

Many emerging health technologies involving digitised patient records, surveillance systems, and biometric data collection raise privacy abuses and concerns about informed consent. These tensions conflict with principles of dignity, autonomy, and communal interests within Ubuntu ethics. For instance, automated patient information systems and centralised databases increase risks of personal health data exploitation by companies, governments, or cybercriminals, often without meaningful patient consent, which violates individual integrity from an Ubuntu lens. Additionally, public health surveillance methods like infectious disease contact tracing, genomics research, and population screening amplify these risks across communities without considering group impacts and tend towards institutional efficiency over participatory governance—contradicting the collective solidarity ethic of Ubuntu (Reviglio & Alunge, 2020).

Ubuntu underscores that any technology mediating health data must uphold individual and collective rights. This demands consent processes recognising relational personhood, minimally intrusive designs limiting unnecessary access, strong protections against exploitation of intimate data, and democratic data governance through inclusive public oversight bodies guided by humanistic values—not just bureaucratic prerogatives. Realising the full benefits of emerging health technologies in equitable, just and empowering ways requires aligning technical design, policy frameworks and cooperative norms with the dignity, relational communitarian principles and social justice foundations at the core of Ubuntu ethics. This provides vital guidance to develop precision public health while avoiding Orwellian pitfalls.

Ubuntu philosophy's communitarian ethics and social justice principles also provide an essential lens for critically assessing specific technology-mediated healthcare approaches that risk eroding professional capabilities and caring bonds between patients, communities, and health workers over time. For example, excessive reliance on automated diagnostic and treatment algorithms may undermine local healthcare professionals' clinical knowledge, judgment and hands-on skills through deskilling effects (Aquino et al., 2023). Eroding individual abilities also impair community healing capacities, which contradicts Ubuntu ethics.

Similarly, automation technologies like robot surgery that optimise narrow efficiency parameters can dehumanise medical care by removing essential human empathy and compassion. Replacing healthcare relationships and social bonds with technology alone violates the Ubuntu ideal of mutual caring. Overall, an Ubuntu analysis reveals the need to holistically evaluate the integration of emerging health technologies against criteria of preserving capabilities, avoiding exploitation, and sustaining reciprocal compassionate relationships that uphold dignity. Reforms must be grounded in the humanistic African ethical perspective. While some technologies augment care, processes shaped overwhelmingly by commercial efficiency paradigms or bureaucratic rationalisation without ethical moorings pose risks of impairing professional skills, moral foundations and social fabrics of healthcare.

Contrasting perspectives on integrating health technologies also emerge when comparing profit-driven private sector approaches focused on commercial returns versus public systems aiming to democratise access according to Ubuntu's communitarian principles. From an Ubuntu ethics viewpoint, private sector healthcare models that ration access to expensive medical technologies according to ability to pay appear deeply unjust as they violate egalitarian ideals of collective care and compassion. The profit motive also incentivises ethically dubious practices like over-diagnosis and unnecessary treatments contradicting patients' needs. Conversely, public health systems designed to provide universal quality care better embody the inclusionary spirit of Ubuntu, even if undermined by state resource constraints and bureaucracies (Jecker et al., 2022).

The humanistic perspective of Ubuntu underscores that healthcare technology innovation and integration must be guided by the fundamental principles of democratising access, compassionate care, and dignity for all—not commercial interests, shareholder returns or administrative self-preservation. This likely requires reforms to prevailing institutional cultures, incentives, and structures across both sectors. Realising equitable, ethical healthcare futures through judicious technology adoption ultimately requires centring the public interest and cooperative values embodied in the African ethics of Ubuntu, not private gain or impersonal institutional agendas. This demands innovation ecosystems guided by concern for collective advancement.

The Ubuntu perspective emphasises the importance of designing and implementing solutions prioritising community engagement and empowerment. Building upon the principles outlined in Fig. 7.1, such as nurturing human bonds, expanding capabilities, and promoting social justice, health technologies should be developed through participatory processes involving local communities in shaping their healthcare solutions. This approach aligns with the Ubuntu value of centring local agencies and priorities, as it recognises the wisdom and expertise of community members in identifying their unique health needs and challenges. By fostering collaborative partnerships between healthcare providers, technology developers, and communities, we can create health technologies that are more culturally sensitive, contextually relevant, and socially empowering.

Furthermore, the Ubuntu ethic of compassion and care for the vulnerable compels us to prioritise the development of health technologies that address the needs of marginalised and underserved populations. This includes designing technologies that are accessible, affordable, and user-friendly for individuals with limited resources, low literacy levels, or physical disabilities. For example, mobile health applications that use simple visual interfaces and local languages can help bridge the digital divide and enable more inclusive access to healthcare information and services. Telemedicine platforms incorporating culturally appropriate communication styles and respect for traditional healing practices can enhance trust and engagement among diverse communities.

7.5   Digital Technology and Ubuntu Ethics

The exponential growth of digital technologies like mobile Internet and social media platforms across Africa has far outpaced ethical analyses to holistically evaluate and steer their effects towards supporting communal well-being—as an Ubuntu ethics perspective would demand. While bringing convenience, growing immersion in virtual environments has also disrupted social relations and awareness. For instance, excessive smartphone use isolates individuals, contradicting Ubuntu's communitarian values. Moreover, unregulated digital spaces enable misinformation, hate speech, and exploitative marketing that violate the public good (Gwagwa et al., 2022). Meanwhile, lacking strong data protections, digital systems are vulnerable to dignity-violating surveillance overreach by corporations and governments. Overall, the Ubuntu ethic underscores the need for innovations to be designed and governed to safeguard collective interests and social fabric beyond merely advancing consumption or control agendas. However, grassroots technology projects like community cellular networks and digital cooperatives guided by Ubuntu principles are more responsibly expanding local connectivity and capabilities. Non-profit and open-access platforms likewise should aim to serve marginalised users rather than mainly extractive profit motives. Such approaches thoughtfully embed technologies within social life.

The growth of automation, artificial intelligence, and digitisation across many spheres raises concerns over the displacement of livelihoods and the erosion of meaning from work, which can be viewed through the ethical lens of Ubuntu philosophy. While improving productivity and administrative efficiency, blind automation contradicts Ubuntu's communitarian and social welfare tenets. For instance, digitising government services risks excluding the less tech-savvy from meaningful bureaucratic work and accessing public benefits. Similarly, AI and algorithms threaten to replace teachers, nurses and other roles providing communal value (Friedman, 2023). Though new jobs may emerge, deskilling risks and lost dignity during transitions violate Ubuntu ethics. Overall, an Ubuntu perspective demands that the impacts of technology changes be centred on worker dignity, capabilities, and livelihoods—not just efficiency. Responsible innovation guided by Ubuntu necessitates balancing productivity gains with social justice through inclusive job transition programmes. More fundamentally, some argue that driving automation and analytics without regard for African relational traditions disrespects humanity and collective advancement (Jecker et al., 2022). Thus, Ubuntu thought presses for cautious, empowering adoption of digital automation. Policy and institutional reforms are imperative to democratise digitisation's advantages while safeguarding work as a dignified vocation. Ubuntu provides wisdom to uphold labour's sanctity amid disruptive change.

The massive growth in digital data extraction and surveillance technologies raises privacy threats from the ethical perspective of the Ubuntu philosophy, which recognises relational personhood and collective interests. As digital profiling and tracking expands through innovations like facial recognition, social media monitoring, and predictive analytics, individual and communal rights face escalating risks of unwarranted access to intimate details by corporate and government entities, enabled by weak consent protections (Reviglio & Alunge, 2020). This violates principles of dignity and integrity from an Ubuntu lens. Thus, Ubuntu presses for urgent reforms to assert data sovereignty as a collective, communal right rather than one transactionally signed away by isolated individuals. Valid consent and negotiating appropriate data sharing require meaningful public debate and consensus (Gwagwa et al., 2022). Rather than unaccountable data extraction, Ubuntu upholds cooperative governance of digital domains guided by transparency, care and democratisation—advancing shared interests, not just institutional efficiency or profits. One model is 'data communities', where citizens collectively control and direct data use, countering external exploitation. Fundamentally, Ubuntu's relational worldview provides grounding for more ethical data agency paradigms that recognise individual and communal personhood. This perspective must shape the evolution of emerging digital ecosystems to protect human dignity and affective ties threatened by unfettered digitisation.

The spread of foreign social media, content platforms, and digital advertising also raises ethical concerns from an Ubuntu perspective regarding risks to cultural self-determination, wisdom transmission and communal values in Africa. The incursion of entertainment and marketing media reflecting Western secular individualism risks gradually eroding traditional African communitarian ethics, localised identities, and intergenerational knowledge exchange centred on oral histories and apprenticeship. Unmoderated digital content also amplifies the risks of youth exploitation by forces seeking cultural control rather than empowerment. However, grassroots initiatives leveraging participatory media, crowdsourced histories, and immersive museum exhibits to sustain cultural heritage digitally and learning show that technologies can be harnessed to strengthen Ubuntu principles of collective memory and identity when conscious intention guides design (Friedman, 2023). Still, policy reforms and educational initiatives are imperative to protect against cultural homogenisation. The relational worldview of Ubuntu ethics provides grounding to consciously shape locally rooted technologies that amplify pluralism and sustain communal values—not impose foreign paradigms eroding traditions. Reformists must balance openness to change with anchoring in indigenous wisdom. Promoting African digital sovereignty and literacy is vital to meet this challenge. Overcoming external cultural hegemony requires a communally shared technology agency guided by pan-African solidarity and Ubuntu principles.

Further, Ubuntu ethics provides a framework for addressing the ethical challenges posed by the rapid expansion of data-driven systems and artificial intelligence. As highlighted in Fig. 7.1, the Ubuntu principles of respecting autonomy, avoiding embedded biases, and promoting social justice are particularly relevant in this context. To uphold these values, developing digital technologies that prioritise transparency, accountability, and user control over personal data is crucial. This can be achieved by implementing robust data protection regulations and promoting digital literacy programmes that empower individuals to make informed decisions about their online presence and privacy (Gwagwa et al., 2020).

Furthermore, the Ubuntu notion of human interconnectedness and the collective good plays a pivotal role in shaping a more inclusive and equitable approach to digital innovation. This approach actively promotes the participation of diverse voices, including women, youth, and marginalised communities, in designing and developing digital technologies. We can create technologies that better reflect African societies' needs, values, and aspirations by fostering a more representative and inclusive digital ecosystem. Initiatives that support local digital entrepreneurship and skills development can help build a thriving African digital economy that generates social and economic benefits for all. This underscores the practical application of Ubuntu ethics in shaping digital innovation and promoting African digital sovereignty and literacy.

7.6   Conclusion

This chapter has underscored the vital role of African ethics in providing philosophical foundations to assess and direct technology trajectories for collective advancement, dignity, and justice. The indigenous wisdom of Ubuntu thought offers indispensable guidance for shaping socially responsible innovation grounded in the region's cultural contexts. The chapter has made several vital arguments. First, mainstream technology paradigms based on individualism, instrumentalism, and technocracy contradict the principles of communitarian ethics and social harmony central to Ubuntu philosophy rooted in sub-Saharan cultures. Second, Ubuntu provides a humanistic lens to identify potential harms from technologies disrupting communal ties and collective interests in pursuing productivity and profits. Its relational worldview presses us to consider social and moral impacts on communities. Third, the Ubuntu principles of solidarity, care, and consensus should consciously steer technology integration to avoid inequality, dehumanisation, and ecological damage caused by unreflective paradigms. Ubuntu thought is indispensable for responsible innovation. Fourth, genuine progress requires grounding technology priorities and designs in indigenous knowledge, not imported models. Local agency and wisdom must direct development. Ubuntu underscores that solutions be contextualised to resonate with African values.

In agricultural technologies, the chapter highlighted how an Ubuntu perspective reveals impacts on communal livelihoods, ecological integrity, and social fabrics often overlooked by narrow productivity metrics. Ubuntu ethics demands thoughtful adaptation of agricultural innovations to local contexts through participatory processes centred on self-reliance, inclusivity, and sustainability. Imposing foreign solutions risks dependence and displacement of smallholders. Biotechnologies also require precautionary scrutiny against collective interests. Regarding health technologies, Ubuntu principles spotlight inequities in access, privacy risks of digitised data, and erosion of caring human bonds. Designing for including marginalised groups, democratic health data governance, and preserving compassionate healing relationships is imperative. Private sector profit motives must be balanced with the public good. Ubuntu offers a basis for equitable, empowering health innovation.

In the digital realm, Ubuntu ethics illuminates threats to social cohesion, cultural sovereignty, and collective rights posed by unregulated expansion of foreign platforms. Immersive virtual worlds can nurture atomistic escapism over grounded communal living. Relational personhood is violated by exploitative data extraction without meaningful public oversight. Contextual digital literacy and policy reforms guided by Ubuntu wisdom are vital to upholding shared dignity. African philosophies like Ubuntu carry essential insights for consciously directing digital transformations.

There is a need to elevate African thought as valid ethical knowledge for assessing technologies and directing innovation rather than uncritically relying on Western paradigms. African intellectual heritage has been discounted for too long, favouring foreign concepts imposed through colonisation and globalisation. Dominant assumptions of individualism and instrumentalism fail to align with traditional African values of collective solidarity. Mainstreaming Africa's philosophical foundations like Ubuntu is thus essential to shaping development trajectories rooted in indigenous wisdom, values and interests, not external agendas. Accessing ethical insights from civilisations forged through Africa's lived experiences provides grounding to consciously integrate technologies enhancing dignity, social welfare, and ecological sustainability on the continent's terms. The communitarian ethics and social justice principles embodied in Ubuntu philosophy provide a transformative framework for reforming technology research, business, and policy ecosystems to serve the collective good. Regular public dialogue and inclusive participation in innovation governance, guided by Ubuntu wisdom, are imperative. Cooperative, contextual models of technology development that centre African agency and ground solutions in local knowledge, values, and needs hold immense promise.

REFERENCES

Amugongo, L. M., et al. (2023). Invigorating Ubuntu ethics in AI for healthcare: Enabling equitable care. In *Proceedings of the 2023 ACM conference on fairness, accountability, and transparency* (pp. 583–592). Association for Computing Machinery.

Aquino, Y. S. J., Rogers, W. A., Braunack-Mayer, A., Frazer, H., Win, K. T., Houssami, N., Degeling, C., Semsarian, C., & Carter, S. M. (2023). Utopia versus dystopia: Professional perspectives on the impact of healthcare artificial intelligence on clinical roles and skills. *International Journal of Medical Informatics, 169*, 104903. https://doi.org/10.1016/j.ijmedinf.2022.104903

Friedman, C. (2023). Ethical concerns with replacing human relations with humanoid robots: An Ubuntu perspective. *AI and Ethics, 3*, 527–538. https://doi.org/10.1007/s43681-022-00186-0

Gwagwa, A., Kazim, E., & Hilliard, A. (2022). The role of the African value of Ubuntu in global AI inclusion discourse: A normative ethics perspective. *Patterns, 3*(4), 100462. https://doi.org/10.1016/j.patter.2022.100462

Gwagwa, A., Kraemer-Mbula, E., Rizk, N., Rutenberg, I., & De Beer, J. (2020). Artificial intelligence (AI) deployments in Africa: Benefits, challenges and policy dimensions. *The African Journal of Information and Communication (AJIC), 26*, 1–28. https://doi.org/10.23962/10539/30361

Jecker, N. S., Atuire, C. A., & Kenworthy, N. (2022). Realising Ubuntu in global health: An African approach to global health justice. *Public Health Ethics*. https://doi.org/10.1093/phe/phac022

Kayange, G. M. (2018). Rediscovering individual-based values in Ubuntu virtue ethics: Transforming corporate entities in post-colonial Africa. In R. Oelofsen & K. Abimbola (Eds.), *An African path to a global future* (pp. 107–128). Council for Research in Values and Philosophy.

Kayuni, H. M., & Tambulasi, R. (2012). Ubuntu and corporate social responsibility: The case of selected Malawian organisations. *African Journal of Economic and Management Studies, 3*(1), 64–76.

Koenane, M. L., & Olatunji, C. P. (2017). Is it the end or just the beginning of Ubuntu? Response to Matolino and Kwindingwi in view of Metz's rebuttal. *South African Journal of Philosophy, 36*(2), 263–277.

Lyles, C. R., Nguyen, O. K., Khoong, E., Aguilera, A., & Sarkar, U. (2023). Multi-level determinants of digital health equity: A literature synthesis to advance the field. *Annual Review of Public Health, 44*, 383. https://doi.org/10.1146/annurev-publhealth-071521-023913

Matolino, B., & Kwindingwi, W. (2013). The end of Ubuntu. *South African Journal of Philosophy, 32*(2), 197–205.

Metz, T. (2007). Toward an African moral theory. *Journal of Political Philosophy, 15*(3), 321–341.

Metz, T. (2011). Ubuntu as a moral theory and human rights in South Africa. *African Human Rights Law Journal, 11*(2), 532–559.

Metz, T. (2014). Just the beginning for Ubuntu: Reply to Matolino and Kwindingwi. *South African Journal of Philosophy, 33*(1), 65–72.

Naude, P. (2019). Decolonising knowledge: Can "Ubuntu" ethics save us from coloniality? *Journal of Business Ethics, 159*(1), 23–37.

Peschard, K., & Randeria, S. (2020). 'Keeping seeds in our hands': The rise of seed activism. *The Journal of Peasant Studies, 47*(4), 613–647.

Ramose, M. B. (1999). *African philosophy through Ubuntu*. Mond Books.

Ramose, M. B. (2002). *African philosophy through Ubuntu* (Revised ed.). Mond Books.

Reviglio, U., & Alunge, R. (2020). "I Am Datafied Because We Are Datafied": An Ubuntu perspective on (relational) privacy. *Philosophy and Technology, 33*(4), 595–612.

van Niekerk, J. (2007). In defence of an auto-centric account of Ubuntu. *South African Journal of Philosophy, 26*(4), 364–368.

Waghid, Y. (2014). *African philosophy of education reconsidered: On being human*. Routledge.

Zimunya, C. T., Gwara, J., & Mlambo, O. B. (2015). The feasibility of an Ubuntu ethic in a modernised world. *Journal of African Foreign Affairs, 2*(1/2), 5–26.

CHAPTER 8

The Spirits in the Machine

8.1   Introduction

Contemporary Africa's rich tapestry of spiritual traditions and religious beliefs has long shaped its diverse peoples' worldviews, values, and cultural practices. From the ancestral veneration of the Yoruba to the animistic cosmologies of the Maasai, African spirituality is deeply intertwined with notions of identity, community, and the sacred interconnectedness of all life. These metaphysical frameworks have proved remarkably resilient and adaptive despite centuries of colonial disruption, cultural upheaval, and modernising pressures. However, the rapid pace of technological change in the twenty-first century presents both challenges and opportunities for African spirituality. On the one hand, the globalising force of digital platforms, mass media, and consumer capitalism threatens to erode traditional value systems, communal bonds, and ways of knowing. The extractive logic of industrial development often clashes with African conceptions of the earth as a sacred living entity, leading to ecological devastation and the desecration of hallowed sites. Fears abound that the seductive pull of virtual realities and AI-driven automation will further alienate people from the tangible rhythms of nature, ancestors, and embodied spiritual practice.

At the same time, African innovators and thought leaders are increasingly harnessing technology as a tool for cultural preservation, creative expression, and social transformation grounded in indigenous wisdom. Digital archiving projects are protecting priceless repositories of oral

© The Author(s), under exclusive license to Springer Nature   151
Switzerland AG 2024
Y. Ndasauka, *African Mind, Culture, and Technology*,
https://doi.org/10.1007/978-3-031-62979-2_8

history, folklore, and traditional knowledge for future generations. Social media and mobile platforms are connecting diasporic communities and amplifying voices of resistance against neo-colonial exploitation. And a new generation of tech entrepreneurs is pioneering Afrocentric solutions in areas like fintech, e-commerce, and renewable energy that prioritise communal well-being over individual gain.

This chapter explores the dynamics between African spirituality and technological innovation in the twenty-first century. It begins by laying out the foundations of African spiritual traditions, including core metaphysical principles, cosmological beliefs, and ethical value systems that have long guided social and ecological relations on the continent. It then examines how these belief systems are shaping African responses to the promises and perils of emerging technologies, from the Internet and smartphones to artificial intelligence and gene editing. The heart of the chapter investigates the tensions, conflicts, and potential synergies between African spirituality and tech innovation. It considers how the disruptive forces of globalisation, consumerism, and digital capitalism are impacting traditional cultural practices and belief systems. It explores the ways in which African spiritual principles of communalism, circularity, and reverence for nature could inform more sustainable and equitable approaches to tech development. It highlights inspiring examples of African innovators who are bridging old and new, sacred and profane, to create culturally resonant technologies that serve the greater good. Finally, the chapter concludes by offering a vision for the future of African spirituality in an increasingly technological world. It argues that by staying grounded in the timeless wisdom of ancestors and the motherland's sacred landscapes while embracing the liberating potential of new tools and ideas.

8.2   Foundations of African Spirituality and Religion

African spirituality and religion are deeply rooted in the continent's ancient history, diverse cultures, and relationship with the natural world. Traditional African religion conceives existence as fundamentally interconnected, a holistic worldview in which the material and spiritual realms are intertwined (Mbiti, 1969). This cosmology shapes beliefs about the divine, ancestral spirits, and humanity's place within the greater web of life. Indigenous belief systems across Africa share some common elements, though specific practices and deities vary by ethnic group and region. One central tenet is the belief in a supreme creator god, often conceived as distant and unknowable (Idowu, 1973). This high god is the source of all existence but not the focus of regular worship. Instead, veneration centres on lesser divinities and spirits that act as intermediaries between the supreme being and humanity (Parrinder, 1969). These entities embody natural forces and human values, serving various roles as guardians, tricksters, or ambassadors to the divine.

Another key concept is the pervasive belief in ancestral spirits that maintain an active presence in the lives of their descendants (Ephirim-Donkor, 2011). Ancestors are seen as wise elders who have crossed over into the spirit world but are still intimately connected to the living. They serve as moral guides and protectors of the community. Venerating the ancestors through offerings, libations, and designated rituals is one of the most important sacred duties of the African religion (Opoku, 1978). This solidifies familial and cultural bonds across generations. The natural world itself is imbued with sacred meaning in African cosmology. Mountains, forests, rivers, and animals can be manifestations of divine spirits worthy of reverence (Beier, 1966). Many ethnic groups have totemic relationships with certain creatures like the python or crocodile, viewing them as sacred symbols of the clan (Uchendu, 1965). Rituals around planting and harvest express gratitude to the earth and nature spirits. Rules against harming sacred groves or species demonstrate an ethic of environmental protection woven into indigenous value systems (Dorm-Adzobu et al., 1991).

This deep interface between spirituality and nature reflects the centrality of the land in traditional African life. Mbiti (1969) explains that in African religions, man is not the master in the universe: He is only the centre, the friend, the beneficiary, and the user. For that reason, he has to live in harmony with the universe, obeying the laws of natural, moral, and mystical order. The earth is seen as the source of life, identity and continuity with the ancestors (Opoku, 1978). Within these belief systems, certain individuals are recognised as having special abilities to mediate between the human and spirit worlds. Priests, diviners, herbalists, and traditional healers undergo training and initiation to serve the community (Some, 1995). Their divination, healing, and intercession skills with the ancestors and spirits are highly regarded. Rituals and rites of passage mark important events in the individual and communal life cycle (Zahan, 1979). Beyond engagement with the spirit world, African traditional religion provides a strong foundation for communal values and a way of life.

Moral and ethical codes centre on sustaining social harmony and cosmic balance. Taboos and prohibitions uphold communal standards. Reconciliation practices and restorative justice aim to rehabilitate offenders and reintegrate them into the community (Elechi, 2007). Elders, in particular, are seen as moral authorities and peacemakers responsible for guiding the group in adherence to ancestral tradition (Yankah, 1995). These humanistic values have enabled many African societies to maintain social cohesion and navigate conflicts. Gyekye (1996) argues that the African ethical outlook is grounded in a humanistic ethic. This moral perspective assigns an inherent value to the human being and places him at the centre of the cosmos. The sacred is experienced through uplifting and unifying the community. However, African cultures have also had to grapple with integrating outside influences and adapting to a changing world. The spread of Christianity and Islam across the continent produced dynamic fusions of traditional African spirituality with monotheistic faiths (Mazrui, 1986). Indigenous belief systems did not simply disappear with colonialism but evolved and found expression within new religious frameworks. Many African Christians and Muslims still engage in rituals venerating ancestors and natural spirits alongside attendance at churches and mosques.

The prominent Yoruba religion of Nigeria exemplifies this synthesis. Yoruba cosmology centres on a high god, orisa spirits, and ancestral forces (Idowu, 1962). With the introduction of Islam and Christianity, the Yoruba were able to assimilate the new religions through a creative process in which the old cosmology was not rejected but built upon (Peel, 1968). Across the African diaspora, Yoruba spiritual concepts significantly shaped syncretic faiths like Vodun in Benin and Haiti, Candomblé and Umbanda in Brazil, and Santería in Cuba (Brown, 1991). These hybrid belief systems reflect the dynamism and resilience of African spirituality in the face of cultural upheaval. African Traditional Religion has become a symbol and an expression of the African spirit and worldview, which continues to inform and pervade the lives of many Africans on the continent and in the diaspora in the midst of social change (Opoku, 1978).

African spirituality and religion encompass a rich heritage of beliefs, practices, and value systems integrally linked with the continent's history and ethnic diversity. The central elements shared across indigenous traditions are reverence for supreme and lesser deities, veneration of ancestral spirits, and a conception of the natural world as imbued with sacred meaning. These metaphysical beliefs sustain a holistic worldview of interconnection among the divine, human, and environmental spheres. From this foundation arises a communal way of life that defines the individual through their social responsibilities and relationships. Moral and ethical codes aim to uphold the collective good and cosmic balance through reconciliatory and restorative practices guided by elder wisdom. Dynamic encounters with Islam and Christianity produced syncretic belief systems that incorporate traditional African spirituality within monotheistic frameworks. This is still the case, but Africa has remained with pockets of its religion and spirituality. Mbiti (1969) claims that the religious heritage of Africa is the product of centuries of development. And it belongs to the African people. At the same time, its humanistic values and integrative philosophies offer insights relevant to a world grappling with questions of ethics, identity and ecology in an age of globalisation and technological change.

8.3   PERSPECTIVES ON TECHNOLOGY

African spiritual worldviews shape perspectives on technology that encompass both practical benefits and existential concerns. On the one hand, indigenous knowledge systems have long embraced innovation and recognised the value of tools for improving quality of life. As shown in Chap. 2, traditional craftsmanship in metallurgy, textiles, and architecture demonstrates advanced technical skills honed over generations. Archaeologists have uncovered evidence of pre-colonial iron smelting and blacksmithing in West Africa dating back over 2500 years. Further, societies like the Yoruba and Benin developed sophisticated urban planning and artistic traditions that required specialised production methods. In this sense, technology is not inherently antithetical to African cultural values when applied towards communal well-being. However, the unprecedented pace and scale of technological change wrought by colonialism and globalisation has provoked deep uncertainty and resistance. Many African societies have experienced technology as a disruptive force imposed from outside rather than emerging organically from their own cultures (Mavhunga, 2017). The abrupt influx of Western technologies during the colonial era was inextricably tied to the experience of foreign domination, exploitation of African land and labour, and suppression of indigenous ways of life.

As a result, its association with imperial conquest often colours attitudes towards technology and an assault on African identity. Certain innovations like railroads and telegraph lines were tools for expanding colonial administration and extracting resources. Christian missionaries used the printing press to spread European languages and literature while denigrating African oral traditions (Maxwell, 2015). The introduction of European medicine and education systems, while bringing some benefits, also undermined indigenous knowledge and inculcated subservience to foreign rule. Given this history, embracing technological progress is not always a straightforward proposition. It raises concerns about the erosion of cultural autonomy and traditional values in the face of a Western-dominated global order. Some African thinkers view modern technology as perpetuating neo-colonial dependency and underdevelopment (see Mazrui, 1986). Reliance on imported technologies and experts can stifle local innovation and entrench power imbalances between Africa and industrialised nations.

From a spiritual perspective, the relentless drive for technological advancement is sometimes seen as a dangerous transgression of natural and sacred boundaries. It represents an attempt to subdue and control the environment without regard for the delicate web of life that sustains humanity's existence (Opoku, 1978). Indigenous ontologies of interconnection counsel restraint and humility in the application of technology rather than the unfettered manipulation of nature for short-term gain. Moreover, some African religious beliefs depict technology as a potential conduit for malevolent supernatural forces. Among the Akan of Ghana, the word for technology translates to 'the devil's boat', implying its corrupting and destructive power (Adas, 1989). Fears that technology can enable witchcraft or open portals to dark spiritual realms are not uncommon. Nwosimiri (2021) notes that Africans are very cautious in their approach to technology because they believe it can be used for pervasive purposes. Rituals of cleansing and protection are thus often invoked around the use of powerful technologies. When factories or large construction projects begin, it is not unusual for traditional priests to perform ceremonies to appease the nature spirits and seek ancestral blessings for a harmonious outcome (see Morgan & Okyere-Manu, 2020). Divination and spiritual consultation may be sought to determine the proper placement and operation of technical systems in alignment with metaphysical forces.

Another significant concern is the ways in which technology can disrupt family bonds and communal ties. African cultural values elevate interpersonal relationships and social solidarity as the essence of human well-being

(Mbiti, 1969). The fear is that excessive dependence on technological devices and systems erodes authentic human connection and collective memory. Turkle (2013) has written extensively about the isolating effects of social media and online interactions replacing in-person contact. In African contexts with strong oral traditions, mass media supplanting face-to-face storytelling and intergenerational knowledge transmission is especially troubling (Peek & Yankah, 2003). Further, rapid urbanisation and globalised consumer culture also threaten the viability of traditional livelihoods and localised economies. The influx of cheap manufactured goods can decimate long-standing handicraft industries and devalue ancestral knowledge. Youth may be drawn away from farming and village life to pursue modern education and urban employment, thereby hollowing out rural communities. While such trends are not unique to Africa, they hold particularly high stakes for cultures still deeply rooted in subsistence agriculture and pastoral life.

Another important consideration is the impact of technology on traditional power structures and gender roles. As discussed previously, access to new technologies can sometimes disrupt long-standing hierarchies and create opportunities for marginalised groups, such as women and youth, to assert greater agency and influence. For example, mobile phones and Internet platforms have enabled women entrepreneurs to bypass traditional gatekeepers and connect directly with markets and resources. However, technology can also exacerbate existing inequalities if access and adoption are uneven across different segments of society.

The role of technology in facilitating cultural exchange and globalisation is another complex issue. On the one hand, digital platforms and media have created new opportunities for Africans to share their stories, arts, and innovations with the world, challenging stereotypes and asserting cultural pride. Social media has also enabled diasporic communities to maintain connections with their homelands and participate in transnational networks of solidarity and activism. On the other hand, the dominance of Western-based tech giants and the homogenising effects of global consumer culture raises concerns about cultural imperialism and the erosion of local traditions and languages.

The increasing importance of data sovereignty and digital security in Africa is another pressing concern. As more African individuals, businesses, and governments embrace digital tools and platforms, vast amounts of sensitive data are being generated and stored, often in the hands of foreign corporations or states. This raises questions about who owns and controls this data, how it is being used, and how to protect it from abuse or exploitation. Efforts to build local data governance frameworks and digital security capacity are crucial for ensuring that African societies can harness the benefits of technology without compromising their autonomy and safety.

The COVID-19 pandemic has underscored both the challenges and opportunities of technology in African contexts. On the one hand, the crisis has exposed the deep digital divides and inequalities within and across African countries, as those without access to connectivity, devices, or electricity struggle to participate in remote learning, work, and social interaction. On the other hand, the pandemic has also accelerated the adoption of digital tools and innovations in areas like health, education, commerce, and governance, creating new possibilities for resilience and transformation. As Africa looks to the future, grappling with the role of technology in building more sustainable, equitable, and resilient societies will be more important than ever.

8.4   Points of Conflict and Reconciliation

Both tension and potential mark the interface between African spiritual traditions and modern technology for synergistic adaptation. At the heart of this dynamic is the question of whether technological advancement undermines or can be harmonised with the humanistic values, communal bonds, and sacred worldviews that have long guided African societies. One fundamental concern is that the proliferation of technology erodes the primacy of human relationships and cultural heritage. African philosophy is grounded in an ethic of interdependence, where individual identity is inseparable from one's social roles and responsibilities to the community. Rituals, oral traditions, and intergenerational knowledge transmission reinforce a sense of collective memory and ancestral continuity. However, the globalisation of information technology and mass media has accelerated a shift towards more individualistic lifestyles and consumption-driven values. Smartphones and social networks can foster a sense of isolation and erode face-to-face interactions (Turkle, 2013). Digital platforms and Western educational models are displacing communal storytelling and traditional knowledge systems. There are fears that African youth are becoming alienated from their cultural roots and that family structures are fracturing under the pressures of urbanisation and technological change.

Another significant tension revolves around the perceived materialism and ecological destructiveness of technology-driven development. As mentioned earlier, many African spiritual traditions view the natural world as imbued with sacred essence, with humans as integral parts of a greater web of life. Reverence for the earth and its bounty is expressed through rituals, taboos, and practices of sustainable resource management passed down over generations. In contrast, the dominant paradigm of techno-logical progress often treats nature as a collection of inert resources to be exploited for human benefit. The extractive logic of industrial capitalism has wrought devastating environmental consequences across Africa. The drive to commodify and financialise every aspect of life threatens the spiri-tual and cultural significance of land, water, and other sacred natural ele-ments. This importation of Western technologies and consumption habits will also exacerbate ecological damage, further alienating people from the life-giving ecosystems they depend on.

The clash between traditional African worldviews and the secularising influence of technology also raises epistemological and existential ques-tions. In many African cultures, spirituality is not a compartmentalised aspect of life but a fundamental way of being and knowing, with the divine manifest in all things. Dreams, visions, and communion with ancestral spirits are considered valid sources of guidance and insight. In contrast, the dominant Western paradigm of scientific rationalism relegates such metaphysical ways of knowing to the realm of superstition or irrationality. This worldview's globalising force, spread through educational systems and popular culture, can create a sense of alienation and disconnection from traditional sources of meaning and identity. It raises questions about the nature of reality, the boundaries of the knowable, and the place of spirituality in an increasingly technoscientific world.

However, while these tensions are real and deeply felt, many points of potential reconciliation and synergy between African spirituality and tech-nological innovation exist. Across the continent, people are finding cre-ative ways to adapt and reappropriate technologies in culturally resonant ways, bridging old and new, sacred and profane. One key principle is the use of technology to support and enhance, rather than replace, Africa's vibrant communal traditions. For example, mobile apps and online plat-forms are enabling African diaspora communities to maintain cultural and linguistic ties with their homelands, organise mutual aid networks, and pool resources for entrepreneurship and development projects. Digital storytelling initiatives are preserving and disseminating indigenous oral histories, folklore, and knowledge systems for future generations.

Another point of reconciliation lies in harnessing technology for environmental stewardship and sustainable development grounded in African values. Across the continent, eco-entrepreneurs and grassroots innovators are pioneering culturally relevant solutions in renewable energy, regenerative agriculture, and circular economies. For example, the Ghanaian startup Recycle Up! trains local artisans to upcycle discarded plastic and e-waste into eco-friendly consumer products, reducing pollution and creating green jobs. In the realm of agriculture, numerous projects are using mobile technologies and data analytics to support smallholder farmers in adopting indigenous permaculture techniques, improving soil health and crop yields while preserving. These examples point to the potential for a uniquely African approach to sustainable development that marries technological innovation with traditional ecological knowledge and ethics of communal care.

Perhaps the integration of African spirituality and technology invites a re-envisioning of innovation as a sacred act of co-creation with the divine. In many African cosmologies, the material and spiritual realms are seen as interconnected aspects of a unified whole, with all phenomena imbued with vital energy and consciousness. From this perspective, engaging with technology is not a purely utilitarian endeavour but a metaphysically and morally charged one, requiring deep reflection, ritual attunement, and responsibility to the greater community of life. This understanding opens up possibilities for intentionally sacralising the process of technological development, infusing it with spiritual awareness and ethical grounding. African innovators and entrepreneurs could draw on ancestral invocation, libation, and divination practices to seek guidance and blessings for their ventures, ensuring that their creations serve the greater good.

The role of African spirituality in shaping ethical frameworks for technology development and governance is an important point of reconciliation. Many African spiritual traditions emphasise values of unity, reciprocity, and reverence for life that could inform more holistic and socially responsible approaches to innovation. For example, the African philosophy of vitalism, which holds that all beings are imbued with a shared life force, could inspire technologies that prioritise ecological balance and regeneration. Similarly, the African emphasis on ancestor veneration and intergenerational wisdom could guide the development of AI systems that respect the knowledge and values of elders and communities. Africans can contribute to a more inclusive and pluralistic vision of progress by bringing African spiritual and ethical perspectives to bear on technological debates.

8.5   INTEGRATING TECHNOLOGY WITH SPIRITUAL VALUES

While tensions exist between African spirituality and modern technology, there are also significant opportunities for harmonising innovation with the continent's cultural and ethical heritage. By proactively shaping technological development through the lens of indigenous wisdom, Africans can craft digitally enabled futures that uplift rather than undermine their deepest-held values. One key principle is leveraging technology to support and enhance Africa's vibrant communal traditions. Extended family networks and kinship ties have formed the backbone of social and economic life for centuries. Rituals, festivals, and rites of passage strengthen intergenerational bonds and cultural continuity. While some fear that digital tools could atomise these social fabrics, they can also be harnessed to facilitate community-building in new ways. For example, as already shown, mobile apps and online platforms connect African diaspora populations and enable them to maintain ties with their ancestral heritage. By augmenting rather than replacing offline interactions, digital technologies can reinvigorate a sense of African identity and solidarity on a global scale.

The strong emphasis on elder wisdom and ancestral guidance in African spirituality also points to ways that technology could bridge generational divides. In many traditions, elders and griots are revered as carriers of communal memory, transmitting knowledge through storytelling, music, and embodied practices. As the old African proverb says, 'when an old man dies, it's a library burning'. Africa can encourage initiatives where elders record their stories and teachings through audio and video, which are then accessible in local languages via mobile apps. This fusion of vernacular expression with digital archiving may help to demonstrate how technology can extend rather than erase traditional modes of intergenerational knowledge transfer. Artificial intelligence can also potentially revalorise African ancestral wisdom in new forms. For example, training natural language processing tools on African orature creates fascinating possibilities for using AI to generate new creative works that build on deep cultural foundations. Another crucial intersection point is harnessing technology to support environmental stewardship and sustainable resource management. Many African cultures have long held nature as sacred, with elaborate rituals and taboos to maintain ecological balance. Rivers, mountains, animals, and plants are often seen as imbued with spiritual essence. Reconciling these attitudes of reverence and restraint with the extractive pressures of industrial technology is an ongoing challenge.

However, indigenous conservation practices are increasingly being augmented with digital monitoring tools and data-driven approaches. For instance, combining scientific data with local knowledge of medicinal plants, animal migration routes and sacred sites, scientists would be better advocates for policies that sustain livelihoods and biodiversity. Across Africa, startups and NGOs are pioneering culturally grounded applications of clean energy, regenerative agriculture, and circular economies. These holistic solutions harness information technology to advance the pan-African ethic of interconnectedness and reciprocity between people and the planet. By reframing technology as a tool for collective well-being and environmental justice, Africans can shape digitalisation to serve the common good.

Embracing technology's spiritual dimensions also means reconceiving it as more than a means to utilitarian ends. Many African philosophies view the cosmos as a unified field of material and immaterial forces, with all phenomena—rocks, plants, animals, ancestors, and deities—participating in this vital energy. Matter and spirit, science and faith, are not opposing poles but part of an integrated whole. In this light, engaging with technology can be re-imagined as a sacred act, an invocation of creative power requiring ethical grounding and ritual attunement. Just as traditional African healers undergo intensive apprenticeship to learn how to wield medicinal and spiritual forces responsibly, African technologists must also cultivate a sense of moral obligation and metaphysical awareness. Weaving spiritual practices like libations, offerings, and meditations into the tech development process would be a powerful way to sacralise it intentionally. Invoking the wisdom of the ancestors through prayer and divination before making significant design decisions, dedicating new products to the service of community well-being, and honouring the hidden energies within our devices are all examples of ritualised tech practices.

Integrating African spirituality with technological innovation is about asserting cultural agency and self-determination. It means shifting from a mindset of passive technology adoption to proactive adaptation aligned with enduring values. However, this does not imply a wholesale rejection of external influences or a romanticised return to pre-colonial ways of life. Rather, it is about critically and creatively engaging with the opportunities of the digital age to advance African aspirations on African terms. It is about cultivating the wisdom to discern what technologies are worth embracing and the courage to reshape or refuse those that threaten essential sources of meaning and identity. As Africa stands on the cusp of a fourth industrial revolution, with rapid advancements in AI, robotics, biotechnology and beyond, these questions of cultural compatibility and spiritual grounding are more pressing than ever. Will the dizzying pace of innovation enrich human existence, or accelerate processes of cultural homogenisation and alienation? Will the benefits of the digital economy be equitably shared, or will they concentrate power in the hands of neo-colonial elites? Will Silicon Valley ideologies bulldoze African worldviews, or will they help steer technology towards more holistic notions of progress?

The potential of virtual and augmented reality technologies to create immersive experiences of African spiritual traditions and sacred sites is an exciting area of exploration. For example, digital reconstructions of ancient African temples, shrines, and pilgrimage routes could allow people to engage with these sacred spaces in new ways, even if they are physically distant or inaccessible. VR simulations of rituals, ceremonies, and initiations could also provide powerful cultural education and transmission tools, particularly for younger generations and diasporic communities. However, the development of these technologies would need to be guided by principles of cultural sensitivity, respect for sacred knowledge, and community ownership and control.

The role of African spiritual leaders and institutions in shaping the development and governance of technology is another important consideration. In many African contexts, religious authorities and structures influence social norms, values, and behaviours significantly. Engaging these leaders as stakeholders, essential knowledge brokers and partners in technology initiatives could help to ensure that innovations are culturally relevant, ethically grounded, and socially accepted (see Ndasauka & Kainja, 2024). For example, collaborations between tech developers and traditional healers could yield new digital health and wellness approaches that integrate spiritual and biomedical knowledge. Similarly, partnerships between religious institutions and digital literacy programmes could help to promote responsible and discerning use of technology, particularly among youth.

The potential of technology to support the documentation, preservation, and promotion of African spiritual heritage is another key area of integration. Many African spiritual traditions rely on oral transmission and embodied practices that are vulnerable to loss and erosion in the face of social and environmental change. Digital tools and platforms could play a crucial role in recording, archiving, and sharing this intangible heritage for future generations. For example, online databases and repositories of spiritual texts, teachings, and artefacts could provide valuable resources for scholars, practitioners, and the general public. Social media and digital storytelling could also help amplify African spiritual communities' voices and experiences, challenging stereotypes and fostering intercultural understanding.

The importance of cultivating spiritual and moral leadership in the African tech sector is another critical point of integration. As the influence and impact of technology in African societies grows, it is essential that those who design, develop, and deploy these tools are guided by strong ethical principles and a sense of social responsibility. This requires technical skills, deep self-awareness, empathy, and wisdom. Integrating African spiritual practices and teachings into tech education and professional development could help foster these qualities and create a new generation of tech leaders who are rooted in their cultural heritage and committed to using their skills for the greater good.

The potential of African spiritual philosophies to inspire new paradigms of technology design and innovation is an exciting frontier of exploration. Many African spiritual traditions emphasise holism, circularity, and regeneration values that challenge Western industrial capitalism's linear, extractive logic. These perspectives could inform the development of more ecologically sustainable, socially equitable, and spiritually nourishing technologies. For example, the African concept of Ubuntu, which emphasises the interdependence of all beings, could inspire the design of collaborative platforms and sharing economies that prioritise mutual aid and collective well-being over individual gain. Similarly, the African reverence for ancestors and future generations could guide the development of long-term, intergenerational technological stewardship and governance approaches.

Driving this conversation is not a matter of pitting tradition against modernity but of drawing on the past to consciously create desired futures. Africa's rich spiritual heritage is not a fossilised relic but a living source of ethical guidance and cultural resilience that can help navigate the promises and perils of technological change. Africa can forge its path between technological determinism and traditionalist stagnation by respectfully adapting ancestral knowledge to contemporary contexts, critically adopting digital tools that serve the common good, and intentionally sacralising innovation processes. It can build digitally enabled societies that are both globally connected and spiritually rooted, using the best of the past and present to create more abundant and meaningful futures for all.

8.6   Conclusion

As we have seen throughout this chapter, the intersection of African spirituality and technological innovation in the twenty-first century is a complex and dynamic space fraught with both perils and possibilities. On one hand, the relentless march of globalisation, digitalisation, and consumerism threatens to erode the very foundations of traditional African belief systems, practices, and ways of life. The individualistic and materialistic values promoted by Western modernity stand in stark contrast to the communal, reciprocal, and spiritually grounded ethics that have long guided social and ecological relations on the continent. As more and more Africans, especially youth, become seduced by the glamour of virtual realities, social media, and the 24/7 news cycle, there is a real risk of losing touch with the tangible rhythms of nature, ancestors, and embodied rituals that have sustained African cultures for millennia.

Moreover, the extractive logic of industrial capitalism, which treats the earth as a collection of resources to be exploited rather than a sacred living entity, has already wrought devastating consequences across Africa. From oil spills in the Niger Delta to mineral conflicts in the Congo, the voracious appetite of global markets has fuelled ecological destruction, social fragmentation, and the desecration of holy sites. The spectre of climate change, driven in large part by the greenhouse gas emissions of the industrialised world, looms as an existential threat to the livelihoods, landscapes, and ancestral traditions of African peoples. Against this backdrop, it is understandable that many Africans view Western technology as a neocolonial imposition, a Trojan horse for cultural imperialism and economic domination. However, to paint African spirituality and technological innovation as inherently incompatible would be a gross oversimplification. As we have seen, African cultures have a long history of creative adaptation and resilience in the face of change. Africans have always been ingenious innovators, harnessing the tools and techniques of their time to serve the needs of their communities. Today, a new generation of African inventors, entrepreneurs, and visionaries is rising to the challenge of shaping technological development in accordance with indigenous values and aspirations.

These trends suggest that, far from being a threat to African spirituality, technological innovation can be a powerful ally in preserving, promoting, and adapting indigenous knowledge and values for the present and future challenges. By harnessing the tools of the digital age to document oral histories, map sacred sites, and connect diasporic communities, African cultures can ensure that their rich spiritual heritage is not lost but rather transmitted and transformed for generations to come. By infusing cutting-edge technologies with the wisdom of ancestors, the sacredness of nature, and the imperatives of social justice, African innovators can pioneer new models of development that prioritise the well-being of people and the planet over the relentless pursuit of profit.

As we conclude this exploration of African spirituality and technological innovation, it is clear that there are no easy answers or one-size-fits-all solutions. The challenges facing Africa in the twenty-first century are formidable, and the forces of globalisation, climate change, and technological disruption show no signs of abating. Yet, in the face of these challenges, African cultures have demonstrated time and again their capacity for resilience, creativity, and adaptive ingenuity. By drawing on the deep wells of spiritual wisdom and social solidarity that have sustained them through countless trials and transformations while embracing the liberating potential of new technologies and ideas, Africans can survive and thrive in the tumultuous decades ahead.

The fate of African spirituality in the age of innovation will depend on the choices and actions of Africans themselves. Will they succumb to the siren song of Western consumerism and individualism, trading the richness of their heritage for the fleeting rewards of modernity? Or will they find ways to harness the tools of technology to preserve, promote, and adapt their indigenous knowledge and values for the present and future challenges? Will they let the forces of globalisation and extraction continue to erode the social and ecological fabric of their communities, or will they pioneer new models of development that prioritise the well-being of people and the planet over the pursuit of profit? These are the questions that Africans in the twenty-first century must grapple with, not only for their own sake but also for the sake of a world in desperate need of new visions and paradigms.

REFERENCES

Adas, M. (1989). *Machines as the measure of men: Science, technology, and ideologies of Western dominance.* Cornell University Press.

Beier, U. (Ed.). (1966). *The origin of life and death: African creation myths.* Heinemann Educational Books.

Brown, D. H. (1991). *Umbanda: Religion and politics in Urban Brazil.* Columbia University Press.

Dorm-Adzobu, C., Ampadu-Agyei, O., & Veit, P. G. (1991). *Religious beliefs and environmental protection: The Malshegu sacred grove in northern Ghana*. Acts Press, African Centre for Technology Studies.

Elechi, O. O. (2007). *Doing justice without the state: The Afikpo (Ehugbo) Nigeria model*. Routledge.

Ephirim-Donkor, A. (2011). *African spirituality: On becoming ancestors* (Revised ed.). UPA.

Gyekye, K. (1996). *African cultural values: An introduction*. Sankofa Publishing Co.

Idowu, E. B. (1962). *Olodumare god in Yorùbá belief*. Longman.

Idowu, E. B. (1973). *African traditional religion: A definition*. SCM Press Ltd.

Mavhunga, C. (Ed.). (2017). *What do science, technology, and innovation mean from Africa?* MIT Press.

Maxwell, D. (2015). The missionary movement in African and world history: Mission sources and religious encounter. *The Historical Journal, 58*(4), 901–930.

Mazrui, A. A. (1986). *The Africans: A triple heritage*. Little, Brown.

Mbiti, J. S. (1969). *African religions and philosophy*. Heinemann.

Morgan, S., & Okyere-Manu, B. (2020). The belief in and veneration of ancestors in Akan traditional thought: Finding values for human well-being. *Alternation Special Edition, 30*, 11–31.

Ndasauka, Y., & Kainja, J. (2024). Stewards or manipulators? Knowledge brokers' complex positionality in combating the COVID-19 infodemic in Malawi. *African Journalism Studies*, 1–17. https://doi.org/10.1080/2374367 0.2024.2341923

Nwosimiri, O. (2021). African cultural values, practices and modern technology. In B. D. Okyere-Manu (Ed.), *African values, ethics, and technology*. Palgrave Macmillan. https://doi.org/10.1007/978-3-030-70550-3_6

Opoku, K. A. (1978). *West African traditional religion*. FEP International Private Limited.

Parrinder, G. (1969). *Religion in Africa*. Praeger.

Peek, P. M., & Yankah, K. (Eds.). (2003). *African folklore: An Encyclopedia*. Routledge.

Peel, J. D. Y. (1968). *Aladura: A religious movement among the Yoruba*. Oxford University Press.

Some, M. P. (1995). *Of water and the spirit: Ritual, magic, and initiation in the life of an African shaman*. Penguin.

Turkle, S. (2013). *Alone together: Why we expect more from technology and less from each other*. Basic Books.

Uchendu, V. C. (1965). *The Igbo of Southeast Nigeria*. Holt, Rinehart and Winston.

Yankah, K. (1995). *Speaking for the chief: Okyeame and the politics of Akan oratory*. Indiana University Press.

Zahan, D. (1979). In T. K. Ezra & L. Martin (Eds.), *The religion, spirituality, and thought of traditional Africa*. University of Chicago Press.

CHAPTER 9

African Philosophy and the Quest for Just Technologies

9.1 INTRODUCTION

In the book *African Mind, Culture, and Technology: Philosophical Perspectives*, I embarked on a multidisciplinary exploration of the relation between African philosophical traditions, cultural dynamics, and technological innovation. Drawing on insights from diverse scholars, thinkers, and practitioners, this volume has sought to illustrate how African epistemologies, values, and lived experiences shape and are shaped by the rapid technological transformations unfolding across the continent and beyond.

Throughout the preceding chapters, I have grappled with fundamental questions about the nature of knowledge, reality, and human existence in an increasingly technologically mediated world. I have examined how African conceptions of personhood, community, and the sacred intersect with the design, adoption, and governance of emerging technologies, from artificial intelligence and biotechnology to digital platforms and renewable energy systems. By engaging with diverse philosophical perspectives from across the continent, this book has challenged dominant Western narratives that often portray Africa as a passive recipient of imported technological solutions rather than a vibrant source of indigenous innovation and creativity. The chapters have highlighted the rich tapestry of African knowledge systems, cultural practices, and ethical frameworks that offer valuable resources for navigating the challenges and opportunities of technological change.

Y. Ndasauka, *African Mind, Culture, and Technology*, https://doi.org/10.1007/978-3-031-62979-2_9

The book has not avoided confronting the tensions, contradictions, and power asymmetries characterising African engagements with technology. The legacy of colonialism, the ongoing realities of economic inequality and political marginalisation, and the threats of cultural homogenisation and environmental degradation all loom large in discussions of Africa's technological future. Yet, amid these challenges, the chapters have also revealed inspiring examples of African ingenuity, resilience, and agency in harnessing technology for social transformation and self-determination. From grassroots innovations in sustainable agriculture and community-based renewable energy projects to digital platforms for preserving indigenous languages and cultural heritage, African innovators are demonstrating that another way is possible: one that puts people and the planet at the centre of technological progress.

As we reflect on the insights and arguments advanced in this volume, several key themes emerge that underscore the significance of African philosophical perspectives in understanding and shaping technological futures, both on the continent and globally. First, the book highlights the importance of grounding technological innovation in African communities' lived realities, cultural contexts, and epistemological frameworks. Rather than imposing one-size-fits-all solutions from outside, the chapters argue for developing contextually relevant technologies that build on local knowledge, priorities, and capacities. This requires a fundamental shift from a technocratic, top-down approach to innovation towards a more participatory, bottom-up model that centres African users' and stakeholders' agency and expertise.

Second, the book emphasises the centrality of ethical considerations in technology design, deployment, and governance. Drawing on African moral philosophies such as Ubuntu, the chapters foreground values such as human dignity, social solidarity, ecological responsibility, and intergenerational justice as critical principles for evaluating the impact of technological interventions. They challenge the notion of technology as a neutral tool and instead reveal its implications for questions of power, equity, and cultural identity. Third, the book points to the urgent need for decolonising and democratising the production and dissemination of technological knowledge. The chapters highlight how dominant global innovation systems often marginalise or appropriate African intellectual property, cultural heritage, and indigenous knowledge. They call for a radical restructuring of the political economy of technology that prioritises

African ownership, control, and benefit-sharing in developing and deploying new technologies.

Finally, the book offers a powerful vision of how African philosophical wisdom can contribute to more humane, sustainable, and socially just technological futures for the continent and the world. By bringing African perspectives to bear on global debates around the ethics of artificial intelligence, the governance of the Internet, the sustainability of energy systems, and the future of work, among other issues, the chapters open up new horizons for imagining alternative pathways of technological progress.

Realising the transformative potential of African insights for shaping inclusive and equitable technological futures will require ongoing multidisciplinary research, dialogue, and action. Building bridges between academia, industry, government, and civil society will necessitate fostering collaborative platforms for knowledge sharing, policy engagement, and innovation practice. Most importantly, it will require centring African communities' voices, experiences, and aspirations, particularly those historically marginalised or excluded from the benefits of technological progress. Only by grounding our efforts in the lived realities and cultural contexts of African peoples can we hope to harness the power of technology for genuine human flourishing and planetary well-being.

In the following sections, we will delve deeper into the specific philosophical dimensions and practical implications of the insights gathered in this volume. We will explore how African epistemologies and metaphysics can inform the design and governance of emerging technologies, how African socio-cultural and ethical frameworks can guide responsible innovation, and how African creativity and spirituality can inspire new visions of technological progress. My goal is not only to contribute to academic knowledge production but also to catalyse transformative action towards more inclusive, sustainable, and culturally resonant technological futures for Africa and beyond. I invite readers to join us on this journey of reflection, dialogue, and co-creation as we work together to build a world where technology serves the needs and aspirations of all people in harmony with the planet and with the deep wisdom of African traditions.

9.2   Epistemological and Metaphysical Dimensions

Exploring epistemological and metaphysical dimensions forms a crucial foundation for understanding the interplay between African philosophical traditions and technological innovation. As highlighted in the introductory chapter, African knowledge systems and conceptions of reality have often been marginalised or dismissed in dominant Western discourses on technology and development. However, the chapters in this volume demonstrate that engaging with African epistemologies and metaphysics is not only valuable but also essential for contextualising and shaping technological futures that are culturally resonant, socially just, and ecologically sustainable.

One of the key arguments advanced in the book is the need to challenge the hegemony of Western scientific rationalism as the sole judge of technological knowledge and progress. Chapter 5 demonstrates that the dominant paradigm of technological determinism, which posits technology as an autonomous force that evolves according to its internal logic, fails to account for how social, cultural, and political factors shape technology's design, adoption, and impact. The chapter calls for a more situated and context-specific understanding of technology that recognises the role of human agency, cultural values, and power relations in shaping technological trajectories. This argument is further elaborated in Chap. 6, which explores how African philosophies of communalism and Ubuntu can inform alternative approaches to technological innovation that prioritise social solidarity, collective well-being, and ecological harmony over individual profit and efficiency. The chapter highlights examples of grassroots innovations, such as community-based renewable energy projects and participatory technology design that demonstrate the potential for African epistemologies to guide more inclusive and sustainable forms of technological development.

The book does not romanticise or essentialise African knowledge systems as monolithic or static. Chapter 2 provides a nuanced account of the diverse and dynamic ways African societies have developed and adapted technologies, from pre-colonial ironworking and textile production to contemporary digital platforms and biotechnologies. The chapter challenges the notion of Africa as a technologically backward or primitive continent. Instead, it reveals a rich history of indigenous innovation and creativity that colonial and neo-colonial powers have often obscured or appropriated. This theme is further explored in Chap. 8, which examines the intersections between African spirituality, technology, and modernity. The chapter argues that African cosmologies and belief systems, which often emphasise the interconnectedness of the material and spiritual worlds, can offer valuable insights for navigating the ethical and existential challenges posed by emerging technologies such as artificial intelligence and biotechnology. Rather than dismissing African spirituality as irrelevant or incompatible with technological progress, the chapter suggests that engaging with these traditions can help to re-centre questions of meaning, purpose, and values in innovation discourses.

Another central argument of the book is that African epistemologies and metaphysics are not antithetical to scientific and technological knowledge but offer complementary and corrective perspectives that can enrich and transform dominant paradigms. Chapter 3 draws on 'epistemological pluralism' to argue the need to integrate African ways of knowing, such as oral traditions, embodied practices, and indigenous ecological knowledge, with modern scientific methods in designing and deploying digital technologies. The chapter highlights projects that have successfully bridged these epistemological divides, such as using mobile apps to document and share traditional medicinal knowledge or the application of data analytics to support smallholder farmers in adopting agroecological practices.

However, the book acknowledges the significant challenges and tensions in integrating African knowledge systems with modern technological frameworks. Chapter 4 examines how the rapid influx of new technologies can disrupt traditional social structures, cultural practices, and value systems, leading to alienation, dislocation, and loss of identity among some African communities. The chapter argues for the need to develop more culturally responsive and contextually appropriate approaches to adoption and adaptation that respect and build upon local knowledge, preferences, and capacities rather than imposing one-size-fits-all solutions from outside. This argument is further elaborated in Chap. 7, which explores how the African ethical framework of Ubuntu can guide the development and governance of technologies in ways that prioritise human dignity, social solidarity, and ecological sustainability. The chapter argues that Ubuntu's emphasis on all beings' intrinsic worth and interconnectedness provides a powerful counterpoint to the instrumentalist and extractivist logic that often drives technological innovation under capitalism. By grounding technology in the values of care, compassion, and reciprocity, Ubuntu offers a vision of progress that is not measured solely in economic growth or efficiency but rather in the flourishing of individuals, communities, and the natural world.

The book's exploration of epistemological and metaphysical dimensions underscores the vital importance of engaging with African philosophical traditions in shaping technological futures that are technically sophisticated, culturally meaningful, socially just, and ecologically regenerative. By bringing African ways of knowing and being to bear on questions of technological design, governance, and impact, the chapters in this volume open up new horizons for imagining and enacting alternative pathways of innovation that centre African communities' agency, dignity, and wisdom. Realising the transformative potential of African epistemologies and metaphysics for technology will require more than just academic theorising. It will necessitate building bridges between different knowledge systems, fostering collaborative partnerships between diverse stakeholders, and investing in the capacity of African institutions and communities to lead and own their technological destinies. It will also require a fundamental rethinking of the political economy of technology, which challenges the concentration of power and resources in the hands of a few and promotes more equitable and inclusive models of innovation and ownership.

9.3   Socio-cultural and Ethical Implications

This book has explored the socio-cultural and ethical implications that form a critical lens to examine the relation between African philosophical traditions and technology. As highlighted in the preceding sections, African knowledge systems and conceptions of reality offer valuable resources for contextualising and shaping technological futures that are culturally resonant, socially just, and ecologically sustainable. However, the chapters in this volume also underscore the challenges and tensions that arise when new technologies intersect with the diverse social structures, value systems, and moral frameworks that characterise African societies.

One central theme throughout the book is the need to interrogate how technology can reinforce and disrupt traditional social hierarchies, gender roles, and power relations. Chapter 4 highlights the negotiations when African communities seek to adapt and integrate new technologies into their cultural practices and belief systems. The chapter argues that while technology can provide opportunities for empowerment and social transformation, it can also exacerbate existing inequalities and create new forms of exclusion and marginalisation, particularly for women, youth, and marginalised groups. This argument is further elaborated in Chap. 7, which examines the ethical implications of emerging technologies such as artificial intelligence, biotechnology, and digital surveillance through the lens of African moral philosophy. The chapter draws on the concept of Ubuntu, which emphasises all human beings' inherent dignity and interconnectedness, to argue for a more humanistic and socially responsible approach to technological development. By grounding innovation in the values of care, compassion, and reciprocity, Ubuntu offers a powerful counterpoint to the instrumentalist and individualistic logic that often drives technological progress under capitalism.

The book also acknowledges the significant tensions and contradictions that can arise when African cultural values and practices conflict with the globalising forces of modernity and technological change. Chapter 8 explores the intersections between African spirituality, technology, and secularisation. The chapter argues that African cosmologies and belief systems can offer valuable insights for navigating emerging technologies' existential and moral challenges. However, they can also be seen as incompatible with the scientific rationalism and individualism underpinning dominant narratives of progress and development. This tension is further explored in Chap. 3, which examines how the rapid proliferation of digital technologies is transforming African social and cultural landscapes. The chapter highlights the potential for digital platforms to enhance access to information, education, and economic opportunities, particularly for marginalised communities. However, it raises concerns about the erosion of traditional forms of community, identity, and social cohesion and the risks of digital surveillance, misinformation, and cultural homogenisation.

A recurring theme throughout the book is the need to develop more culturally responsive and contextually appropriate approaches to technology adoption and governance that respect and build upon local values, priorities, and capacities. Chapter 6 draws on the theory of social construction of technology (SCOT) to argue that technologies are not neutral or deterministic but are shaped by the social, cultural, and political contexts in which they are developed and used. The chapter highlights examples of grassroots innovations and participatory design processes demonstrating the potential for African communities to actively shape and adapt technologies to meet their needs and aspirations. This argument is further elaborated in this chapter, which explores the role of African creativity, aesthetics, and cultural expression in shaping technological innovation. The chapter argues that rather than simply importing or imitating Western technology design and production models, African innovators can draw on the continent's rich cultural heritage and artistic traditions to develop more locally relevant and culturally resonant technologies. From using traditional textile patterns in digital interfaces to incorporating oral storytelling in virtual reality experiences, the chapter highlights the potential for African cultural resources to inspire new forms of technological expression and innovation.

The book also acknowledges the significant ethical challenges and risks arising when external actors use African cultural heritage and intellectual property to be appropriated, commodified, or exploited. Chapter 2 provides a critical analysis of how African indigenous knowledge systems and technologies have been marginalised, suppressed, or co-opted by colonial and neo-colonial powers. The chapter argues for the need to decolonise the production and dissemination of technological knowledge and protect African communities' rights to own, control, and benefit from their cultural resources and innovations. This theme is further explored in Chap. 5, which examines the ethical implications of the global politics of technology transfer and the unequal distribution of the benefits and risks of technological development. The chapter argues that African countries and communities must have a more significant say in shaping the governance of emerging technologies, particularly those that significantly impact their livelihoods, environments, and cultural identities. It calls for a more inclusive and participatory approach to technology policy and decision-making that centres the voices and experiences of African stakeholders and challenges the dominant power structures and interests that often drive technological agendas.

9.4   EXISTENTIAL AND SPIRITUAL DIMENSIONS

The book has also examined the existential and spiritual dimensions of technological innovation in Africa. As highlighted in the preceding sections, African philosophical traditions and cultural frameworks offer rich resources for contextualising and shaping technological futures that are more holistic, humanistic, and ecologically attuned. However, the chapters in this volume also underscore the challenges and tensions that arise when emerging technologies intersect with African spiritualities, cosmologies, and ontologies and the need for more integrative and culturally resonant approaches to innovation and development.

One central theme throughout the book is the need to recognise and engage with the deep spiritual and existential dimensions of African worldviews in shaping technological trajectories. Chapter 8 explores the intersections between African spirituality, technology, and modernity. The chapter argues that African cosmologies and belief systems, which often emphasise the interconnectedness of the material and spiritual worlds, can offer valuable insights for navigating the ethical and existential challenges posed by emerging technologies such as artificial intelligence, biotechnology, and virtual reality. Drawing on examples from various African cultures, the chapter demonstrates how African spiritualities can provide a holistic and relational framework for understanding the nature of being, agency, and consciousness in technological mediation. It suggests that engaging with these ontological perspectives can help to re-centre questions of meaning, purpose, and value in innovation discourses and challenge the reductionist and materialist assumptions that often underpin dominant paradigms of technological progress.

The book also acknowledges the significant tensions and contradictions arising when African spiritual traditions encounter the secularising and globalising forces of modernity and technological change. Chapter 4 examines how the rapid influx of new technologies can disrupt traditional cultural practices, social structures, and belief systems, leading to alienation, dislocation, and loss of identity among some African communities. The chapter highlights the risks of cultural homogenisation, commodification, and appropriation that can occur when African spiritual heritage is decontextualised, digitised, or commercialised without proper safeguards and ethical considerations. This theme is further explored in Chap. 7, which examines the ethical implications of emerging technologies through the lens of African moral philosophy. The chapter argues that the dominant paradigms of technological innovation, which often prioritise efficiency, control, and individualism over communal well-being and ecological balance, are fundamentally at odds with the values of Ubuntu, which emphasise the inherent dignity, interdependence, and sacredness of all life forms. It calls for a radical reorientation of innovation agendas towards more holistic, compassionate, and spiritually grounded visions of progress that affirm all beings' unity and the natural world's intrinsic worth.

Indeed, a recurring theme throughout the book is the need to develop more integrative and culturally resonant approaches to technological innovation that bridge the divide between the material and the spiritual, the scientific and the sacred, the local and the global. Chapter 3 draws on "techno-animism" to explore how African spiritualities and ontologies can inform the design and use of digital technologies in more culturally meaningful and ecologically sustainable ways. The chapter presents case studies of African digital artworks, virtual reality experiences, and video games incorporating elements of African mythology, cosmology, and ritual practice to create immersive and transformative encounters with the sacred. These examples demonstrate the potential for African spiritual traditions to inspire new forms of technological expression and innovation that challenge the binary oppositions between nature and culture, mind and matter, self and others that often characterise Western thought. They suggest that by re-enchanting technology with a sense of wonder, reverence, and reciprocity, African innovators can create more holistic and life-affirming visions of the future that honour all beings' interconnectedness and the earth's sacredness.

The book also recognises the significant challenges and risks of integrating African spiritualities and technologies in global capitalism and neo-colonial power relations. Chapter 5 examines the ethical and political implications of Western corporations and institutions' commodification and appropriation of African spiritual knowledge and cultural heritage. It argues that the uncritical adoption of foreign technological systems and paradigms can lead to the erosion of African ontological frameworks, epistemological traditions, and value systems and the imposition of a homogenised and secularised worldview that undermines the diversity and resilience of African cultures.

9.5   Transforming Innovation Ecosystems and Policies

The book has also explored strategies for transforming innovation ecosystems and policies that harness African philosophical insights to guide more inclusive, sustainable, and culturally resonant approaches to technological development. Realising this transformative potential will require fundamentally restructuring the institutional, economic, and political landscapes that shape technological trajectories in Africa and beyond.

One of the key arguments advanced in the book is the need to decolonise and democratise the production and governance of technological knowledge in Africa. Chapter 5 provides a critical analysis of how the dominant paradigms of science, technology, and innovation (STI) have been shaped by the legacies of colonialism, racism, and epistemic violence. The chapter argues that the uncritical adoption of Western scientific methods, technological systems, and innovation models has marginalised African ways of knowing and doing and perpetuated dependency, exploitation, and underdevelopment relations. The chapter calls for a radical decolonisation of African STI institutions, curricula, and research agendas to address these challenges. This involves increasing the representation and leadership of African scholars, innovators, and communities in these spaces and fundamentally reorienting the epistemological and ontological foundations of knowledge production and technological design. It means centring African languages, histories, and philosophies in STI education and research and valorising indigenous knowledge systems and local innovation practices as legitimate sources of expertise and creativity. This argument is further elaborated in this chapter, which explores the role of African cultural heritage and artistic traditions in inspiring new forms of technological expression and innovation. The chapter highlights the importance of protecting and promoting African intellectual property rights, traditional knowledge, and cultural resources in the context of global innovation systems that often appropriate and commercialise these assets without proper recognition or benefit-sharing. It calls for more equitable and culturally responsive approaches to innovation that build on African communities' strengths and aspirations. It prioritises the agency and self-determination of African creators and innovators.

Another key theme throughout the book is the need to nurture endogenous innovation capabilities and infrastructures in Africa. Chapter 3 examines how the interests and agendas of foreign corporations, development agencies, and state actors have shaped the rapid proliferation of digital technologies in Africa. While acknowledging the potential benefits of digital platforms and services for enhancing access to information, markets, and public services, the chapter also highlights the risks of digital colonialism, surveillance capitalism, and technological dependency that can arise when African countries rely heavily on imported technologies and expertise. The chapter calls for greater investment in African technology hubs, maker spaces, and innovation ecosystems to mitigate these risks and promote more locally driven and contextually relevant innovation. This involves providing financial and technical support to African entrepreneurs, startups, and SMEs and fostering collaborative networks and partnerships between academia, industry, government, and civil society. It means creating enabling policy environments that incentivise local research and development, technology transfer, and capacity building, as well as prioritising African users' and communities' needs and aspirations. This theme is further elaborated in Chap. 6, highlighting the importance of participatory and inclusive innovation processes that engage diverse stakeholders in the co-design, co-production, and co-governance of technological solutions. The chapter presents case studies of grassroots innovation movements that demonstrate the potential for bottom-up, context-specific, and socially embedded approaches to innovation that challenge the top-down, expert-driven, and market-oriented models that often dominate in Africa.

Transforming innovation ecosystems and policies in Africa will require a multifaceted and collaborative approach that engages diverse actors and sectors. For example, the African Union's Science, Technology and Innovation Strategy for Africa 2024 (STISA-2024) emphasises the need for a more integrated and coordinated approach to STI policy and governance across the continent, one that aligns with the aspirations of the African Union's Agenda 2063 for inclusive and sustainable development. The strategy calls for greater investment in research and development, technology transfer, capacity building, and promoting regional and international cooperation in STI. Similarly, the United Nations' 2030 Agenda for Sustainable Development and its associated Sustainable Development Goals (SDGs) provide a global framework for innovation that prioritises social, economic, and environmental sustainability. The SDGs emphasise the importance of inclusive and equitable innovation that leaves no one behind and addresses the needs and challenges of the most vulnerable and marginalised communities. They also call for greater collaboration and partnership between governments, businesses, civil society, and academia in pursuing sustainable development.

In Africa, scholars and practitioners emphasise the need for more context-specific and culturally responsive approaches to innovation that build on the strengths and aspirations of local communities. For example, 'innovation for inclusive development' (IID) has emerged as a critical framework for understanding and promoting innovation that benefits marginalised and underserved populations. IID emphasises the importance of participatory and collaborative innovation processes that engage local communities in the co-design and co-production solutions that address their specific needs and challenges. Similarly, the concept of 'frugal innovation' has gained traction as a way of understanding and promoting innovation that is affordable, accessible, and appropriate for resource-constrained environments. Frugal innovation involves developing products and services that are low-cost, high-quality, and locally relevant, often by leveraging local resources and knowledge systems. Examples of frugal innovation in Africa include developing mobile money services, solar-powered irrigation systems, and low-cost medical devices.

Transforming innovation ecosystems and policies in Africa will require a fundamental shift in the values, priorities, and power relations that shape technological development on the continent. It will require greater recognition of African communities' agency, creativity, and wisdom in driving their innovation agendas and a more equitable and inclusive approach to innovation governance that prioritises the needs and aspirations of the most marginalised and vulnerable populations. This will involve reforming existing institutions and policies and creating new spaces and platforms for dialogue, collaboration, and co-creation between diverse actors and sectors. It will require greater investment in research and development, technology transfer, capacity building, and promoting regional and international cooperation in STI. It will also need a more holistic and integrated approach to innovation that considers technological change's social, cultural, and environmental dimensions and prioritises the well-being of people and the planet over short-term economic gains.

9.6   Conclusion

In conclusion, this chapter has synthesised critical insights from the book *African Mind, Culture, and Technology: Philosophical Perspectives* to propose policy, practical, and ethical frameworks for shaping technology to promote human development in Africa. Drawing on the communitarian values articulated throughout the volume, the chapter has demonstrated that African philosophical traditions offer vital conceptual resources for creating more just, empowering, and culturally resonant technological futures. The chapter has highlighted the importance of grounding innovation in African epistemologies, ontologies, and value systems, emphasising the interconnectedness of all beings, the centrality of community and social solidarity, and the inherent dignity and worth of every person. It has proposed specific principles rooted in Ubuntu and Afrocentric thought that innovators and policymakers should apply to elevate technology to enhance collective well-being rather than exacerbate inequality, alienation, or environmental destruction.

To realise this vision, the chapter has outlined strategies for building endogenous scientific, engineering, and innovation capacities in Africa to increase African agency and self-determination in shaping technological trajectories. This involves decolonising technology by designing systems that harmonise with, rather than disrupt, African social life, cultural heritage, and ecological landscapes. It also requires transforming innovation ecosystems and policies to prioritise inclusive participation, equitable benefit-sharing, and responsible stewardship of technological resources. The chapter has also presented practical models for the participatory development of technologies tailored to community needs, aspirations, and values. These include grassroots innovation movements, maker spaces, and collaborative platforms that engage diverse stakeholders in the co-design, co-production, and co-governance of locally relevant solutions. By centring the voices, experiences, and creativity of African communities, particularly those historically marginalised, these approaches can help make innovation participatory and ensure its benefits are more widely and equitably distributed.

The chapter concludes that applying African wisdom traditions to science, technology, and innovation can make these domains more participatory, restorative of human dignity, and conducive to sustainable development. By grounding technological progress in the timeless values of Ubuntu, such as compassion, reciprocity, and harmony with nature, African societies can chart alternative pathways of modernity that eschew the destructive excesses of Western industrialism and individualism. As humanity grapples with the existential threats posed by climate change, rising inequality, and the unchecked power of artificial intelligence, African philosophical perspectives offer timely and compelling resources for re-imagining the relationship between technology and society. By engaging seriously with these perspectives and working to translate them into concrete policies, practices, and designs, we can chart a more hopeful and humane course for the future of technology, not just in Africa but also for the world.

INDEX

MIX
Papier aus verantwortungsvollen Quellen
Paper from responsible sources
FSC® C105338
FSC
www.fsc.org

Printed by Libri Plureos GmbH
in Hamburg, Germany